A. Marneros N. C. Andreasen
M. T. Tsuang (Eds.)

Psychotic Continuum

Contributors
N.C. Andreasen, J. Angst, F.M. Benes, R.W. Buchanan
W.T. Carpenter Jr, T.J. Crow, A. Deister, M. Flaum
J.A. Fleming, B. Kirkpatrick, A. Marneros, M. Martin
H.Y. Meltzer, Ch. Mundt, H. Remschmidt, A. Rohde
E. Schulz, J.C. Simpson, G.-E. Trott, M.T. Tsuang
D.P. van Kammen

With 29 Figures and 23 Tables

Springer-Verlag
Berlin Heidelberg New York London Paris
Tokyo Hong Kong Barcelona Budapest

Prof. Dr. med. Andreas Marneros
Martin-Luther-Universität Halle-Wittenberg
Klinik und Poliklinik für Psychiatrie
06097 Halle, Germany

Prof. Nancy C. Andreasen, MD, Ph.D
Department of Psychiatry
University of Iowa College of Medicine
MHCRC, 2911, JPP, 200 Hawkins Dr.
Iowa City, IA 52242, USA

Prof. Ming T. Tsuang, MD, Ph.D, D.Sc
Department of Psychiatry
Harvard Medical School at Brockton/West Roxbury
VA Medical Center,
940 Belmont Street
Brockton, MA 02401, USA

ISBN-13: 978-3-642-79487-2 e-ISBN-13: 978-3-642-79485-8
DOI:10.1007/ 978-3-642-79485-8

Library of Congress Cataloging-in-Publication Data. Schizophrenia: psychotic continuum or distinct entities/ A. Marneros, N. C. Andreasen, M. T. Tsuang, editors. p. cm. Includes bibliographical references and index. ISBN-13: 978-3-642-79487-2 (New York) 1. Schizophrenia – Classification. 2. Schizophrenia Diagnosis. I. Marneros, A., (Andreas), 1946– . II. Andreasen, Nancy C. III. Tsuang, Ming T., 1931– . [DNLM: 1. Schizophrenia – pathology. WM 203 S337964 1995] RC514.S33627 1995 DNLM/ DLC for Library of Congress 94-47536

Softcover reprint of the hardcover 1st edition 1995

Typesetting: Best-set Typesetter Ltd., Hong Kong

SPIN: 10480202 25/3130/SPS – 5 4 3 2 1 0 – Printed on acid-free paper

Preface

One of the most important questions of our previous common volumes about affective, schizoaffective, and schizophrenic disorders was the question of what connects and what separates psychotic disorders (Marneros and Tsuang, *Schizo-affective Psychoses*, Springer-Verlag, 1986; Marneros and Tsuang *Affective and Schizoaffective Disorders*, Springer-Verlag, 1990; Marneros, Andreasen, and Tsuang, *Negative and Positive Schizophrenia*, Springer-Verlag 1993).

The boundaries between various psychotic disorders are not always clearly defined. Some groups of psychotic disorders, such as schizoaffective disorders and all the other "atypical" psychoses, occupy a position between "typical" mental disorders, such as schizophrenia, and affective disorders. The question is: Do psychotic disorders form a continuum, or are they, despite their unclear boundaries, distinct entities?

On what basis should we assume there is a continuity of psychotic disorders? Solely symptomatology? Or perhaps also a continuity in genetic predispositions? Or in biological, pharmacological, and other dimensions? Is the old idea of "Einheitspsychose" (unitary psychosis) really always wrong?

The contributions contained in this new volume cannot provide a definite answer to the above questions. But they try to describe some relevant aspects of the problem, and to give some partial answers.

Halle-Wittenberg, Germany
Brockton, USA
Iowa, USA
October 1994

A. Marneros
M. T. Tsuang
N. C. Andreasen

Contents

Part I

Part II

List of Contributors

NANCY C. ANDREASEN, M.D., Ph.D.
The Department of Psychiatry, The University of Iowa College of Medicine, MHCRC, 2911 JPP, 200 Hawkins Drive, Iowa City, IA 52242, USA

Prof. Dr. med. JULES ANGST
Psychiatrische Universitätsklinik Zürich, Postfach 68, 8029 Zürich 8, Switzerland

FRANCINE M. BENES, M.D., Ph.D.
Harvard Medical School, McLean Hospital, 115 Mill Street, Belmont, MA 02178-9106, USA

ROBERT W. BUCHANAN, M.D.
Maryland Psychiatric Research Center, Department of Psychiatry, University of Maryland School of Medicine, Baltimore, MD 21226, USA

WILLIAM T. CARPENTER JR., M.D., Ph.D.
Maryland Psychiatric Research Center, Department of Psychiatry, University of Maryland School of Medicine, Baltimore, MD 21226, USA

TIM J. CROW, M.D.
University Department of Psychiatry, Warneford Hospital, Oxford OX3 7JX, UK

Priv. Doz. Dr. A. DEISTER
Klinik und Poliklinik für Psychiatrie der Universität Bonn, Sigmund-Freud-Straße 25, 53127 Bonn, Germany

Dr. MICHAEL FLAUM
The Department of Psychiatry, The University of Iowa College of Medicine, 200 Hawkins Drive, Iowa City, IA 52242, USA

JEROME A. FLEMING, M.D.
Department of Psychiatry, Harvard Medical School at Massachusetts Mental Health Center, 74 Fenwood Road, Boston, MA 02115, USA

Brian Kirkpatrick, M.D.
Maryland Psychiatric Research Center, Department of Psychiatry,
University of Maryland School of Medicine, Baltimore, MD 21226, USA

Prof. Dr. med. Andreas Marneros
Klinik und Poliklinik für Psychiatrie der Martin-Luther-Universität
Halle-Wittenberg, 06097 Halle/Saale, Germany

M. Martin
Klinik für Kinder- und Jugendpsychiatrie der Philipps-Universität Marburg,
Hans-Sachs-Straße 6, 35039 Marburg, Germany

Herbert Y. Meltzer, M.D.
Department of Psychiatry, University Hospitals of Cleveland, Case Western
Reserve University, Hanna Pavilion, Room B-68, 2040 Abington Road,
Cleveland, OH 44106-5000, USA

Prof. Dr. med. Ch. Mundt
Psychiatrische Klinik der Ruprecht-Karls-Universität Heidelberg, Voßstraße 4,
69115 Heidelberg, Germany

Prof. Dr. med. Dr. phil. Helmut Remschmidt
Klinik für Kinder- und Jugendpsychiatrie der Philipps-Universität Marburg,
Hans-Sachs-Straße 6, 35039 Marburg, Germany

Priv. Doz. Dr. med. Anke Rohde
Klinik und Poliklinik für Psychiatrie der Martin-Luther-Universität
Halle-Wittenberg, 06097 Halle/Saale, Germany

E. Schulz
Klinik für Kinder- und Jugendpsychiatrie der Philipps-Universität Marburg,
Hans-Sachs-Straße 6, 35039 Marburg, Germany

Prof. John C. Simpson
Department of Psychiatry, Harvard Medical School at Massachusetts Mental
Health Center, 74 Fenwood Road, Boston, MA 02115, USA

G.-E. Trott
Klinik für Kinder- und Jugendpsychiatrie der Philipps-Universität Marburg,
Hans-Sachs-Straße 6, 35039 Marburg, Germany

Ming T. Tsuang, M.D., Ph.D.
Department of Psychiatry (116 A), Harvard University, Brockton VA Medical
Center, 940 Belmont Street, Brockton, MA 02401, USA

Daniel P. van Kammen, M.D., Ph.D.
Department of Psychiatry, University of Pittsburgh Medical Center,
Highland Drive, Pittsburgh, PA 15206, USA

Part I

Psychotic Continuum: An Introduction

A. Marneros

Approximately 10 years ago Ming Tsuang and I discussed the DSM-III classification. We thought that some diagnostic groups of DSM-III regarding "cases-in-between" of the so-called typical mental disorders involved some problems. It was obvious that we needed much more research and discussion on "cases-in-between." This was one of our intentions when we organized the first European–American workshop on psychotic disorders with the topic schizoaffective disorders in Düsseldorf in 1985 (Marneros and Tsuang 1986). One conclusion from the first workshop was that schizoaffective disorders comprise an inhomogeneous group in between affective and schizophrenic disorders, perhaps related more to affective than to schizophrenic disorders.

At the next workshop (1988) the question was addressed of how strong these relations are between affective and schizoaffective disorders and what the similarities and differences are between them (Marneros and Tsuang 1990). We concluded that there are many more similarities than differences between affective and schizoaffective disorders. One interesting similarity concerns the aspect of the unipolarity or bipolarity of the affectivity. Perhaps there are no affective and no schizoaffective disorders but rather only unipolar and bipolar disorders as two separate but very large groups (Marneros et al. 1991b). The assumption of the French psychiatrists Fallret in 1851 and Baillarger in 1854 and of the German group of Wernicke, Kleist, and Leonhard, namely, that unipolar and bipolar affective disorders are distinct entities, has been confirmed with modern methods and definitions and developed and reformulated by the very important studies of Jules Angst and Carlo Perris in 1966 as well as by the work of George Winokur and Paola Clayton 2 years later (Marneros et al. 1991b).

Some groups, especially the group of Jules Angst and ours, found that the schizoaffective disorders must be divided into unipolar and bipolar in exactly the same way as affective disorders are. These findings make the position of schizoaffective disorders as "disorders-in-between" much more interesting for both theoretical and empirical work.

What is the relation of unipolar and bipolar schizoaffective disorders to schizophrenia? One possible answer could be the continuity model. Heinz Häfner described this model at our second workshop in Luxemburg in 1988 in

Psychiatric University Hospital Halle, 06097 Halle, Germany

the following way: "The continuity concept can be called the 'green disease model,' with schizophrenia being defined as the 'blue disease,' and bipolar affective disorders as the 'yellow disease.' The schizoaffective symptomatology would then be defined the rich green color obtained by mixing the two primary colors representing the major functional psychoses. This suggests that there is a continuum between unipolar, bipolar, schizoaffective, and schizophrenic psychoses. This model would cover all the gradations of color ranging from yellow through yellow-green and from blue through blue-green to green." This raises a naive question, according to Heinz Häfner, namely, continuity of what? In response to Häfner's "naive" question, we thought that it could be very fruitful to first try to determine whether phenomenologically defined schizophrenic subtypes are independent entities or interrelated and unstable phenomenological constellations. We thought that the model of the positive-negative dichotomy of schizophrenia might be a useful model for such a discussion (Marneros and Tsuang 1990).

That was the topic of the third European-American workshop (Marneros et al. 1991a). The conclusions reached at this workshop underlined some of the limitations of such a dichotomy. In particular the longitudinal work presented by Häfner and our group has demonstrated a polymorphism of psychotic disorders in one and the same patient if the illness has a long duration. Tim Crow was very categorical in his contribution, claiming there are no entities or separate diseases but rather a continuum. However, Heinz Häfner's question still has not been answered: a continuum of what?

A very old answer to this evergreen question is the concept of the unitary psychoses. Although this term was invented by K. Jaspers in 1913, the concept itself is a very old one, originating with Neumann, Zeller, Griesinger and others at the beginning and the middle of the last century in Germany and with Chianugi in Italy. In the United Kingdom Jackson can be considered as the great exponent of nineteenth century "unitarianism," as Berrios pointed out (Mundt and Sass 1991).

The idea of unitary psychoses is still alive in Germany. The psychopathological model of Werner Janzarik ist related to such a concept and my predecessor as Psychiatric Chair of the Martin-Luther-University in Halle, Germany, Prof. Hellmut Rennert, postulated between 1964 and 1984 a "universal genesis of psychoses." Even Kurt Schneider's concept has some relation to the unitary concept: Kurt Schneider accepted only differential typologies but no differential diagnoses in psychiatry.

Berrios (in Mundt and Sass 1991) underlined some points related to the concept of unitary psychoses. He wrote:

Although it is difficult to identify the assumptions shared by all those who have declared in favour of the "unitary psychosis," the following provide a start: 1. there is only one psychosis, 2. it results from a disorder in an invariant "structure," 3. clinical "differences" between "psychoses" reflect pathoplastic effects. Unfortunately, these criteria require qualification before they are of use to anyone.

In regard to criterion 1, for example, the meaning of "psychosis" needs clarification: does it refer to the existence of only one disease (in which case the nosological view is accepted?). Does it simply refer to a psychology of dynamic configuration, to a manner of expressing distress (in which case no commitment is made to nosology)? Indeed, this latter view is almost tantamount to saying that there are as many psychoses as there are patients.

Criterion 2 is also obscure: what is the nature of the invariant structure? Is such required even by those who do not hold a nosological view? A perusal of the literature shows that such a structure has been variously conceived as being organic in nature, i.e., as a brain module, circuit, or site clearly defined in space (e.g., Neumann); or as psychological, i.e., as a process stretching in time (e.g., Freud, Devereux); or, indeed, in structural terms, i.e., as a configuration existing both in space and time (e.g., Ey, Llopis and Janzarik).

Finally, criterion 3 is of use only if it offers testable hypotheses: what should be regarded as a real difference between psychoses? What is a pathoplastic effect? Are personality, key precipitants, or life events more relevant in this context? What is the role of perceptual sets and observer bias in the ascertainment of such differences?

Modern ideas of psychotic continuum are not necessarily identical to concepts such as unitary psychosis or universal genesis of psychoses. The modern use of the term "psychotic continuum" is based on empirical data which are interpreted by authors such as Jules Angst in the sense of a continuum of psychopathological groups and in the sense of the impossibility for a clear dichotomy into schizophrenia and affective disorders. Or in the sense of variations, for example, Tim Crow concluded that there is a continuum of psychotic illness including variations with the affective and schizophrenic realms as well as between the major Kraepelinian prototypes (Marneros et al. 1991a).

What can support such views, what can disprove them?

We hope some answers to these and other questions will be found in the contributions in this volume.

References

Marneros A, Tsuang MT (eds) (1986) Schizoaffective psychoses. Springer, Berlin Heidelberg New York

Marneros A, Tsuang MT (eds) (1990) Affective and schizoaffective disorders. Similarities, differences. Springer, Berlin Heidelberg New York

Marneros A, Andreasen N, Tsuang MT (eds) (1991a) Negative versus positive schizophrenia. Springer, Berlin Heidelberg New York

Maraneros A, Deister A, Rohde A (1991b) Affektive, schizoaffektive and schizophreen Psychosen: eine vergleichende Langzeitstudie. Springer, Berlin Heidelberg New York

Mundt C, Sass H (eds) (1991) Für und wider die Einheitspsychose. Thieme, Stuttgart

Psychotic Continuum or Distinct Entities: Perspectives from Psychopathology

Ch. Mundt

Introduction

The pursuit of clear-cut and theoretically convincing classifications of the idiopathic psychoses by means of operationalized psychopathology was instigated 30 years ago in order to carry out epidemiological studies, studies of the long-term-course, and studies of treatment efficacy. Today, the revitalized genetics need sound categorical classification. However, one has to bear in mind that this is but one reason for typing psychiatric patients. A psychotherapist who wants to address coping behavior, or avoidance of precipitating life events, or training methods for deficits will have quite different diagnostic demands; again other aspects of typing may serve the pharmacologist, such as the Kielholz design, which in the meantime is questioned. Thus, the following considerations will leave aside pragmatic aspects of classification, e.g., for treatment, but instead refer to classical nosological research, the principles of which were outlined by Kraepelin (1903) and, in the English scientific world, by Cullen in 1803 (see Berrios and Beer 1992) as follows:

Disease entities are determined by identical symptoms, course, etiology, and pathomorphology. The research process is meant to follow a constant approximation of this ideal. Hence, outlining perspectives from a mere psychopathological point of view is a difficult task, since psychopathological validation of a nosological concept usually implies external validation with course, outcome, biological parameters, genetics, and personality and will probably result in outlining a type of mental disorganization, perhaps inherent to its structure, rather than producing stringent mental concomitants of particular biological disturbances.

With regard to the psychopathological aspects of the four classical categorization compounds suggested by Kraepelin and Cullen, in the following paper I will outline first some representative results of research carried out by means of operationalized psychopathological instruments. Second, I will discuss an example of a psychopathological system which I would like to call trans-symptomatological, because it tries to recognize the more basic psychic mechanisms which lead to the evolvement of manifest symptoms. All symptom orders which suggest a hierarchy have to work with such a trans-symptomato-

Psychiatrische Klinik, Universität Heidelberg, Voß Str. 4, 69115 Heidelberg, Germany

logical hypothesis, be it Jaspers' rule of psychopathological layers (Jaspers 1965), Bleuler's basic and accessory symptoms (Bleuler 1969), or Carpenter's primary and secondary negative symptoms (Carpenter et al. 1985). In conclusion I will give my personal view on the basis of the available results.

The Symptom Level

As far as *cross-sectional analyses of the symptom specificity* are concerned, which used psychopathometric instruments and operationalized diagnoses, several techniques have been applied. One is to assess a large sample of mixed diagnoses from the endogenous spectrum in order to determine whether there is a bimodel or unimodel distribution of symptoms. Furthermore, associations between symptomatology and course – Kraepelin's way – can be tested. Kendell and Brockington (1980) did just that with their classical studies. The first study by Kendell and Gourley (1970) yielded a trimodel distribution which could be considered consistent with a distinction of schizophrenic, schizoaffective, and bipolar affective disorder. Kendell's second study together with Brockington (1980) yielded a unimodel distribution of symptoms over the whole spectrum, with the maximum of cumulative frequency of all symptoms in the schizo-affective middle area.

In conjunction with Kendell's studies, an investigation should be mentioned which was done by H. Kick (1981) of the Heidelberg group, who reassessed diagnoses according to the patients' records with diagnoses which had been made by Kreapelin himself from 1891 to 1903, the period when Kraepelin was chairman of the Heidelberg University Hospital. Kick applied the AMDP system to the psychopathology laid down in the patients' files. To his surprise, it was not possible to discriminate between schizophrenia and mania unless the course during the (at the time, much longer) inpatient period was taken into consideration.

Even if the comparison between diagnostic groups of the spectrum is narrowed down to symptoms which are considered specific for some groups, specificity cannot be proven. Köhler et al. (1979) demonstrated that Schneiderian first-rank symptoms – which Kurt Schneider believed to be pathognomonic for schizophrenia – occurred at least as often in schizoaffective patients as in schizophrenics.

So-called basic symtoms, the actual core of schizophrenia in the view of the Huber group, were found to a high degree in personality disorders, neuroses, and affective disorders, and negative symptoms were found by our group to overlap greatly among different diagnostic groups, among them major depression and different kinds of personality disorders and neuroses (Mundt and Kasper 1987).

Silverstein and Arzt (1985) examined Chapman's theory of excessive yielding in word associations and found no specificity among schizophrenics, schizoaffectives, and manics. It seems that more complex constructs are needed, as in Berner's (1982) Viennese schizoprenia critera, for example, in which "increased drive" rules out scoring thought disorder in the sense of schizophrenic thought disorder. This is due to the fact that manic flight of ideas and schizophrenic loosening of associations are difficult to discriminate in purely linguistic terms.

Moreover, blurred nosological borders on the symptom level are found not only between schizophrenia and bipolar affective disorder but also between schizophrenia and borderline personality disorder and between schizophrenia and organic psychoses. Johnstone et al. (1978) demonstrated the broad overlap between schizophrenia and organic psychoses, as did Kane et al. (1988) more recently. Unspecificity is shown even more if every occurrence of a symptom over the long-term course is scrutinized, such as delusion or severe mood changes.

The same unspecificity can be observed in the borderland between psychopathology and psychophysiology. In the 1970s, Heimann (1986) showed that disturbances of the orienting reaction are found among schizophrenics as well as among depressives. His co-worker Frank Schneider (1992) extended this approach to analyses of mimical reactions and negative contingency waves and again found no specificity for schizophrenia.

A summary of cross-sectional studies on the single-symptom level (see Table 1) shows overwhelming evidence for unspecificity. Therefore, models have been looked for which combine or hierarchize symptoms, in the hope that clinical impressions of better specification can be empirically tested.

One major topic has been symptom-cluster models: Pioneers in this type of research were Lorr and his group (Lorr et al. 1963). Working with the IMPS, Lorr found ten clusters which averaged nine single items each, for example, clusters for delusion, apathy, obsession, or organic state, due to what he called psychopathological first-order vectors. He called those regular symptom aggregations syndromes. They have been reproduced many times, among others by the German system AMDP. The syndromes show up with different loadings in different nosological categories. Lorr called them syndrome profiles. Second-order factor analysis revealed what Lorr called second-order vectors of psychopathology, and they seem to be more specific than the single symptom. With regard to schizophrenic samples, these vectors were apathy vs. excitation, paranoid process, and disorganization. These three principles organized schizoprenic psychopathology in an unselected sample of 380 acute and chronic patients. This finding has been replicated quite frequently since then. Some of

Table 1. Nosological unspecificity in cross-sectional psychopathological studies

- Overall symptoms and outcome correlations
- First-rank symptoms (Schneider)
- Basic symptoms (Huber)
- Negative symptoms (Andreasen)
- Thought disorders (Chapman)
- Psychophysiological orienting reaction (Heimann)

Table 2. Relative specificity of the three-factor model

Reference	Method	Finding
Lorr et al. (1963)	Second-order factor analysis, IMPS acute and subacute patients	Paranoid projection, disorganization, apathy
Mundt (1985)	Cluster-analysis IMPS on chronic patients	Paranoid projection, motor disturbance, apathy
Bilder et al. (1985)	Factor analysis SANS, external validation	Par/hall. syndrome, disorganization, apathy
Liddle et al. (1987)	Factor analysis SEDS, chronic patients	Paranoia, thought disorder, retardation
Mundt (1989)	Order analysis InSKa, chronic patients	Parathymia, thought disorder, autism/asthenia
Landmark et al. (1990)	Factor analysis PSE, WHO cross-cultural data, external validation	Auditory hallucinations, experiences of interference, parathymia

the studies are listed in Table 2. Moreover, with neuroimaging techniques different patterns of disturbed cerebral activity were assigned to the three psychopathological syndromes. Although the results of the studies show quite a remarkable consistency in their results, they are not absolutely congruent. The latent trait of all applied systems as reflected by the results seems to be a combination of alienation, disorganization, and retardation, which fits well with the classical, mainly Bleulerian, interpretation of the illness, i.e., that production and maintenance of structured meaning and purpose, or their disturbance are the essence of schizophrenia as well as the mental sequelae.

However, in all of these studies demarcation of nosological categories by means of comparative classification has not been a major aim. Instead, validation of the adopted psychopathological rating scales was generally the goal. It must be understood that the three-factor model will have discrete borders to other nosological entities. Shifts of the loadings on the different syndromes, and thus a change of the proportions within the syndromes, will characterize transitions to other nosological categories rather than simple presence or absence of symptom clusters.

Although the three-factor model – strictly speaking – does not prove the distinct entity schizophrenia, it does suggest a certain consistency of the clinical observations which are due to the used instruments. The borders of the syndrome are blurred or discrete; the core, however, can be determined by fairly precise statements.

Attempts have been made to establish other factor models. Crow's (1986) and Andreasen and Olsen's (1982) two-factor model of positive and negative schizophrenia does not contribute much to the demarcation problem, although

Barnes et al. (1989) demonstrated that negative symptoms and depressive symptoms do not covariate in the short-term course, which supports the specificity hypothesis. On the other hand, the same can be said about the co-occurrence of negative symptoms and cognitive disturbances, as Harvey and Serper (1990) and our group (Barnett and Mundt 1992) have discovered, which argues against the specificity hypothesis of the negative syndrome. Validation studies concerned the long-term course, and they produced contradictory results. Since the longitudinal aspect will be dealt with by A. Marneros, this aspect will not be considered further here.

A five-factor model was recently assorted by Lindenmayer et al. (1994). It encompasses clusters of positive symptoms, cognitive failures, excitement, retardation, and depression. If the clusters excitement and retardation are taken together as one vector and depression is considered difficult to separate in the scales from negative symptoms, the scheme comes very close to Lorr's. It seems to be statistically sound but more difficult to interpret clinically, and it has not been replicated as yet by another group. In the literature it does not play a major role, in contrast to the two- and three-factor models.

Another attempt to find better classification systems deals with matching vs. non-matching symptoms. The most successful example of matching symptom combinations is the discrimination between mood-congruent and mood-incongruent delusion. If the type of mood and the type of delusion match, delusion is not considered a sign of schizophrenia; if they do not match, delusion is considered indicative. This is the same point Kurt Schneider made with the distinction between paranoid reaction and delusional perception. Again, the validation of such distinctions has to rely on external parameters, usually on the longitudinal aspect and genetics. Tsuang et al. (1990) recently found evidence for the discriminative validity of the symptom combination, as far as long-term course and outcome are concerned. Similar to the three-factor models, the symptom combinations become meaningful in the light of clinical interpretation. Mood-borne and affect-sustained distortions of reality constitution do not necessarily reflect structural frailty of cognition and affective regulation per se, but may represent an epiphenomenon of another type of antecedant disturbance.

The idea of hierarchies of symptoms was first worked out by Jaspers (1965) with the rule of the different nosological strata: neurosis, affective disorder, schizophrenia, and organic disturbances, the last one being the most fundamental. This *Schichtenregel* implies that the more fundamental diagnosis rules out the ones above it in the hierarchy. This model obviously was given up, at least as far as the relation affective psychosis–schizophrenia is concerned, when the schizoaffective psychoses were incorporated into the affective psychoses in DSM-III. It is a horizontal, rather than a vertical hierarchical model. A renaissance of this idea is Fould's and Bedford's hierarchical model. Interesting theoretical analogies of the "layer model" are Jackson's theory of positive symptoms by disinhibition and Henry Ey's theory.

Thus, if we sum up the perspectives of psychopathological research with operationalized rating scales and leave aside the longitudinal aspect, the three-factor model and, to a certain extent, the idea of matching symptoms characterize schizophrenia as a distinct entity, although its borders remain blurred.

Trans-symptomatological or Functional Psychopathology

The expression "trans-symptomatological psychopathology" shall signify approaches which try to recognize functional principles behind the symptoms, principles which are able to explain the multitude of clinical phenomena and can guide further research. These approaches must contain an interpretative element. One early example was Bleuler's (1969) discrimination between primary and secondary symptoms, which implies a functional hypothesis. The isolation of such functions and their pathological disturbance, irrespective of the nosological adherence, was much appreciated in biological psychiatry, when von Praag (von Praag et al. 1987) promoted it with the model of functional deviances in the serotonin system. He suggested a functional psychopathology instead of a nosological one. Functional psychopathology in von Praag's sense can also be pursued by mere psychopathological methods, and it is by no means novel.

Instead of providing an endless and tedious review of such interpretative systems, I will select Janzarik's (1988) structural-dynamic psychopathology as a model of this type of thinking, and I will try to outline the perspectives and fallacies of such an approach. The structural-dynamic approach has the advantage that it sheds some light on the continuum question, compared, for example, with the vulnerability model as another piece of functional psychopathology.

The origins of Janzarik's considerations were equally a clinical observation and contact with a theory which fit the observation well. The theory refers to the so-called second Leipzig Gestalt psychology school and their term "structure", which denominates the value system of a person. The value system contains the most important life goals, internalized duties, anxieties, strivings, motives. The clinical observation concerned the interdependence of dynamic expansion in acute schizophrenia and in schizoaffective states with the value system in form of delusions or hallucinations, which, in Janzarik's view, are overactivated parts of the value system. A side effect of this view led to a completely different definition of hallucinations than that in DSM-III-R. Janzarik does not regard hallucinations as false perceptions, but as free-running contents of the value system. Fragile personality structures and high dynamic charge on a particular knot of the value system contribute to the content's free running being set off. In later years, Janzarik hypothesized a third contributing factor: the inability to extinguish contents which urge themselves into the psychic field and turn them from the intentional mode to the impressive mode, which latter prevails in the delusional mood (*trema sensu*, Conrad 1958). These ideas resemble very much Peter Liddle's claims of supervisory processes concerning the control of activity and the suppression of inappropriate responses. Janzarik's system also implies a developmental psychopathology, one element of which is the assumption of a growing but latent premorbid pathology which prepares for the actual onset.

One of the main criticisms concerns the difficulty in defining and determining personality dynamism. This is more difficult than determining structure, for which cognition and language and personality can serve as models. It can be approached by neuropsychological and linguistic investigation and by evaluation of schizoid and schizotypical personality. But dynamism is seen only in its almalgmation with structure and is thus difficult to operationalize.

According to Janzarik, the two functional elements structure and dynamism can be disturbed separately; this would be equivalent to the poles of the spectrum of idiopathic psychoses. However, what usually happens is derailment of dynamism, which activates knots of the value system, which in turn become free running under the influence of expanded dynamism. The overlap of activated dynamism and altered structure marks, according to Janzarik, is the core of idiopathic psychoses. In consequence, he considers schizoaffective psychosis the actual center of the spectrum; he does not see it as an accident or as falling in between the actual entities, as Kurt Schneider did.

Janzarik's system of structural-dynamic interplay shall serve as just one model of trans-symptomatological psychopathology to overcome nosological futility. The system parallels von Praag's functional psychopathology in what Janzarik used to call pure or mere psychopathology, i.e., psychopathology without reference to other than mental phenomena. The two functional elements structure and dynamism and possibly the ability to extinguish unwanted, overobtrusive mental contents could be, according to Janzarik, separately transmitted genetically, which would account for the endless variations of clinical pictures, given the developmental influences on the almalgamations between dynamism and structure. This system should be made testable for the techniques of modern cognitive neurosience.

Conclusion

On the single-symptom level, no specifity for schizophrenia and thus no single disease entity can be found within the spectrum of the idiopathic psychoses. However, several techniques of symptom combinations have yielded fairly consistent symptom clusters. They signify core syndromes which make sense clinically. The borders of the sydromes remain blurred. It must be assumed that the discrete character of the entities is due not to unsufficient research techniques but rather to their very nature. All operationalized instruments work with the nineteenth-century symptom catalogue which has been put into them. Thus, the results cannot go beyond; at best they support them. Conventions rather than objective findings mark the borders, whereas the extremes are very well discriminable.

Kraeplin's idea of nosology based on psychopathology has been overemphasized by Jaspers and Kurt Schneider. Nineteenth-century eidetic psychopathology cannot satisfy the demands of Kraepelin's nosological ideal. If psychopathology is forced to comply with the postulated categorizations, overexact dogmatic differentiations occur, as in Kurt Schneider's work. They force clinical reality into a straitjacket. The Heidelberg School suffered from this nosological preoccupancy until the era of Janzarik.

However, this incompatibility between Kraepelin's nosological demands and the means of clinical psychopathology does not invalidate the latter. One possible conclusion is that Bonhoeffer's (1910) statements on organic reaction types may serve as a model for idiopathic psychotic reaction types as well. Many different biological and psychosocial influences can contribute to the etiology of the same psychopathological syndrome. Its color can be tinged by one or the other concomitant symptom, but essentially the mind works with the same limited set of symptoms and a few stereotypes of functional disorganization. Therefore, transnosological psychopathology may be an alternative. It should imply both biological functional entities and trans-symptomatological functional psychological entities; it seems that longitudinal aspects, especially those before the onset of the illness, specify better. As far as trans-symptomatological functional models are concerned, I can imagine that research which deals with the organization of neural networks at present is best suited to link psychological and biological functional units to a functional psychopathology beyond nosological preoccupancy.

References

Andreasen NC, Olsen S (1982) Negative vs positive schizophrenia. Definition and validation. Arch Gen Psychiatry 39:789–794

Barnes TRA, Curson DA, Liddle PF, Patel M (1989) The nature and prevalence of depression in chronic schizophrenic in-patients. Br J Psychiatry 154:486–491

Barnett W, Mundt C (1992) Are latent thought disorders the core of negative schizophrenia? In: Spitzer M, Uehlein F, Schwartz MA, Mundt C (eds) Phenomenology, language, and schizophrenia. Springer, Berlin Heidelberg New York, pp 240–257

Berner P (1982) Achsensyndrome endogener Psychosen. In: Huber G (ed) Endogene Psychosen: Diagnostik, Basissymptome und biologische Parameter. Schattauer, Stuttgart, pp 77–96

Berrios GE, Beer D (1992) Unitary psychosis in English speaking psychiatry: a conceptual history. In: Mundt C, Sass H (eds) Für und wider die Einheitspsychose, Thieme, Stuttgart, pp 12–21

Bilder RM, Mukherjee S, Rieder RO, Pandurangi AK (1985) Symptomatic and neuropsychological components of defect states. Schizophr Bull 11:409–419

Bleuler E (1969) Lehrbuch der Psychiatrie, 11th edn. Springer, Berlin Heidelberg New York

Bonhoeffer K (1910) Die symptomatischen Psychosen im Gefolge von akuten Infektionen und inneren Erkrankungen. Deuticke, Leipzig

Carpenter WT, Heinrichs DW, Alphs LD (1985) Treatment of negative symptoms. Schizophr Bull 11:37–49

Conrad K (1958) Die beginnende Schizophrenie. Thieme, Stuttgart

Crow TJ (1986) The continuum of psychosis and its implication for the structure of the gene. Br J Psychiatry 149:419–429

Harvey PC, Serper MR (1990) Linguistic and cognitive failures in schizophrenia. J Nerv Ment Dis 178:487–493
Heimann H (1986) Spezifität und Unspezifität bei psychiatrischen Erkrankungen. Schweiz Arch Neurol Psychiatry 137:67–86
Janzarik W (1988) Strukturdynamische Grundlagen der Psychiatrie. Enke, Stuttgart
Jaspers K (1965) Allgemeine Psychopathologie, 9th edn. Springer, Berlin Heidelberg New York
Johnstone EC, Crow TJ, Frith CD, Camey MWP, Price JS (1978) Mechanisms of the antipsychotic effect in the treatment of acute schizophrenia. Lancet 1:848–851
Kane J, Honigfeld G, Singer J, Meltzer H, the Clozapin collaborative study group. (1988) Clozapin for the treatment-resistant schizophrenic. Arch Gen Psychiatry 45:789–796
Kendell RE, Brockington JF (1980) The identification of disease entities and the relationship between schizophrenia and affective psychoses. Br J Psychiatry 137: 324–331
Kendell RE, Gourley J (1970) The clinical distinction between the affective psychoses and schizophrenia. Br J Psychiatry 117:261–266
Kick H (1981) Die Dichotomie der idiopathischen Psychosen im Syndromprofilvergleich der Kraepelin'schen Krankheitsbeschreibungen. Nervenarzt 52:522–524
Koehler K, Seminario J (1979) Research diagnosable "schizo-affective" disorders in Schneiderian "first rank" schizophrenia. Acta Psychiat Scand 60:347–354
Kraepelin E (1903) Psychiatrie. Ein Lehrbuch für Studierende und Ärzte. Barth, Leipzig
Landmark J, Merskey H, Cernovsky Z, Helmes E (1990) The positive triad of schizophrenic symptoms. Its statistical properties and its relationship to 13 traditional diagnostic symptoms. Br J Psychiatry 156:388–394
Liddle PF (1987) The symptoms of chronic schizophrenia. A re-examination of the positive-negative dichotomy. Br J Psychiatry 151:145–151
Lindenmayer J-P, Bernstein-Hyman R, Grochowski S, Bark N (1995) Psychopathology of schizophrenia: Initial validation of a 5-factor model. Psychopathology 28:22–31
Lorr M, Klett CJ, McNair DM (1963) Syndromes of psychosis. Pergamon, Oxford
Mundt Ch (1985) Das Apathiesyndrom der Schizophrenen. Eine psychopathologische und computertomographische Untersuchung. Springer, Berlin Heidelberg New York
Mundt Ch, Kasper S (1987) Zur Schizophreniespezifität von negativen und Basissymptomen. Nervenarzt 58:489–495
Mundt Ch, Kasper S, Huerkamp M (1989) The diagnostic specificity of negative symptoms and their psychopathological context. Br J Psychiat 155 [Suppl 7]:32–36
Mundt Ch (1989) Die psychopathologischen Grundlagen der Psychopathometrie des schizophrenen Residualsyndroms. Fundam Psychiatr 3:2–11
Schneider F (1992) Psychophysiologische Unspezifität schizophrener Erkrankungen. Fischer, Stuttgart
Silverstein ML, Arzt AT (1985) Neuropsychological dysfunction in schizophrenia. J Nerv Ment Dis 173:341–346
Tsuang MT, Lyons MJ, Faraone SV (1990) Heterogeneity of schizophrenia. Br J Psychiatry 156:17–26
Von Praag HM, Kahn RS, Asnis GM, Wetzler S, Brown SL, Bleich A, Korn ML (1987) Denosologization of biological psychiatry or the specificity of 5-HT disturbances in psychiatric disorders. J Affective Disord 13:1–8

Psychotic Continuum Under Longitudinal Considerations

A. Marneros, A. Rohde, and A. Deister

The debate concerning the controversy of psychotic continuum, or "unitary psychosis," and separate entities is as old as scientific psychiatry itself (Berrios and Beer 1992; Schmidt-Degenhardt 1992; Vliegen 1980). Distinguished psychiatrists in Germany, France, Great Britain, and other European countries supported the idea of unitary psychosis or of a universal pathogenesis of psychoses while other authors supported the idea of independent entities (Vliegen 1980). At the end of the nineteenth and, mainly, the beginning of the twentieth century, however separate entities had become widely accepted. The well-established system known worldwide following the idea of separate entities is Kraepelin's dichotomy of functional disorders in dementia praecox (later schizophrenia) and manic-depressive illness (Kraepelin 1883–1920). Even more consequently than Kraepelin Wernicke (1900), Kleist (1928, 1953) and Leonard (1957, 1980) followed the idea of separate entities in psychiatry. In his *Aufteilung der Endogenen Psychosen* (Classification of Endogenous Psychoses) Leonard (1957) presented a taxonomy, like his teacher Kleist, of endogenous psychoses with a lot of groups and subgroups. Such a plethora of disease entities, in spite of some fruitful ideas, is connected with a lot of problems. Even followers, or part followers, of that concept found in empirical work that the boundaries between some of Leonard's diseases (for instance, among the various subtypes of the "cycloid psychoses") are very unclear (Perris and Brockington 1981). Additionally longitudinal research did not support the existence of some of the described disease entities, such as "pure mania" (Angst 1966; Perris 1966; Marneros et al. 1988, 1989, 1991). Patients with a monomorphous manic course of disease over many years are so seldom (perhaps no more than 5%) that it can be assumed that a pure manic course is only an artifact (Angst 1966, 1978; Perris 1966; Marneros et al. 1991c).

In contrast to Leonhard's taxonomy, Kraepelin's classification is really simple and clear. Some of the problems connected with Kraepelin's taxonomy, however, are related directly to the simplicity of the system. An example of this is Kraepelin's consideration of unipolar and bipolar disorders as a single disease. The elimination of Fallret's (1851) concept of "*folie circulaire*" or of Baillarger's (1854) concept of "*folie à double forme*" is not only an example of the simplicity but also of the inhibiting role of this system (Angst 1966;

Psychiatric University Hospital Halle, 06097 Halle, Germany

Perris 1966; Marneros 1988, 1989, 1991). The practicability and clearness of Kraepelin's concept, in spite of those and other problems, were important reasons for its relatively good acceptance and even survival, but even Kraepelin had a lot of doubts and problems with his own system. In his very important publication of 1920 *Die Erscheinungsformen des Irreseins* (The Manifestations of Insanity) he expressed his doubts: "Perhaps it is also possible to tackle the difficulties which prevent us from distinguishing reliably between manic deny that there is an alarmingly large number of cases in which it seems impossible, in spite of the most careful observation, to make a firm diagnosis. satisfactorily between these two illnesses and this brings home the suspicion that our formulation of the problem may be incorrect."

Nevertheless, the modern debate concerning the antinomy "psychotic continuum and distinct entities" concentrates mainly on Kraepelin's dichotomy, perhaps because it is practicable yet clear. The most important elements of Kraepelin's bimodal taxonomy were onset, course, and outcome of mental diseases. The question now is: how much do we know about those three aspects a century later and after the creation of an International Classification of Diseases with ten revisions, and after the triumph of DSM-III and at the beginning of the era of DSM-V?

In this paper we present some data from our studies and discuss them in correlation to the literature regarding psychotic continuum. The data have been selected from the Cologne Study (Marneros et al. 1991a). We are not going to present population, methods, definitions etc. because we have already presented them repeatedly in our earlier European-American workshops and have published them in several papers (Marneros and Tsuang 1986, 1990; Marneros et al. 1990, 1991a–c).

Onset

Age of Onset

The first element of Kraepelin's distinction of dementia praecox (schizophrenia) from manic-depressive insanity (affective disorders) was the age of onset: Schizophrenia is correlated with early onset and manic-depressive illness with an onset later in life.

We could confirm this (Fig. 1). The average age of onset in schizophrenia is 27.7 years, in affective disorders 36.1; the difference is statistically significant. But this separation is very simple and only true if we ignore some important details. There are, namely, some intermediate stages or "bridges" between the two sides, if we differentiate between unipolar and bipolar forms and if we do not ignore the schizoaffective disorders, as Fig. 2 shows.

Phenomenology of Syndromes at Onset

According to the bimodal concept we could expect only three types (perhaps exceptionally four types) of phenomenological constellations of "endogenous"

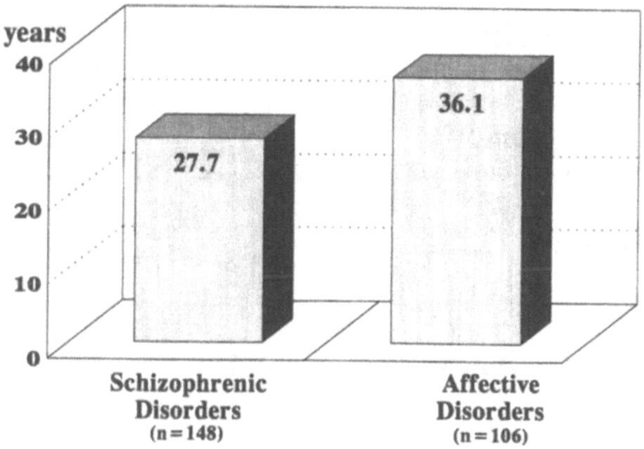

Fig. 1. Age at onset. Schizophrenia versus affective disorders

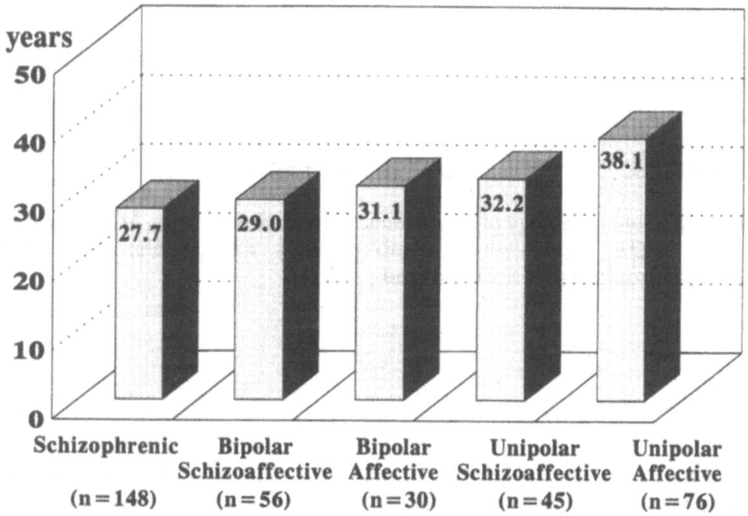

Fig. 2. Age at onset considering the "intermediate" stages

psychoses, namely, a schizophrenic phenomenological constellation or a melancholic, a manic and perhaps – only exceptionally – a manic-depressive-mixed episode (according to Kraepelin (1899) and his pupil Weygand a "mixed state"), as is demonstrated in Fig. 3.

The clinical reality, though, does not conform with these expectations, but rather with a psychopathological picture such as that presented in Fig. 4. The phenomenological constellations of the onset of "endogenous" psychoses are really much more complicated, with at least seven types of episodes: schizophrenic, melancholic, manic, manicdepressive-mixed, schizodepressive, schizomanic and schizomanic-depressive-mixed episodes. That means we often have a combination of schizophrenic and affective symptomatology.

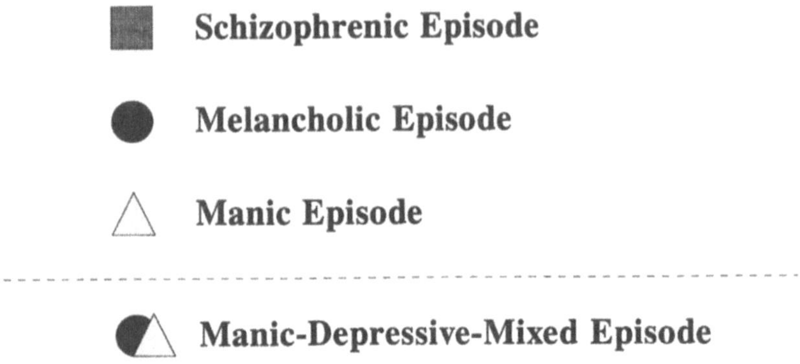

Fig. 3. Pure syndrome manifestation: hypothetical expectations

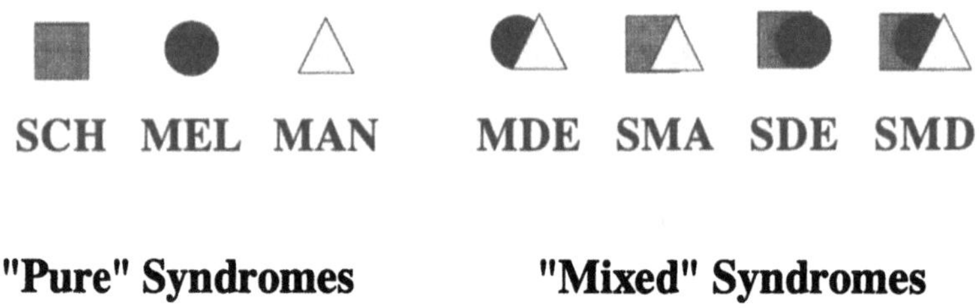

Fig. 4. Cross-sectional phenomenology of syndromes. *SCH*, schizophrenic; *MEL*, melancholic; *MAN*, manic; *MDE*, manic-depressive mixed; *SMA*, schizomanic; *SDE*, schizodepressive; *SMD*, schizomanic-depressive-mixed

Course

If we accept that mental disorders are absolutely distinct and independent entities, then we have to expect a strictly *monomorphous course*, i.e., only one and the same type of episode during the whole course can be manifested (as Fig. 5 demonstrates in theory). The clinical reality, however, is different, as Fig. 6 shows, presenting the results regarding course after the initial schizophrenic episode.

Although there is not a straight line of longitudinality, meaning there is not a monomorphous course, we still have good longitudinal stability: 92% of the patients beginning with schizophrenic episodes have *only* schizophrenic episodes in their course, but 8% of them have other types of episodes (these findings are compatible with findings in the literature). However, there is one detail regarding the longitudinal stability of schizophrenic disorders which is not unimportant: The reported, good longitudinal stability of schizophrenic syndromes in fact concerns the whole phenomenological spectrum of schizophrenic syndromes, i.e., all subtypes of schizophrenia. Within this spectrum one type

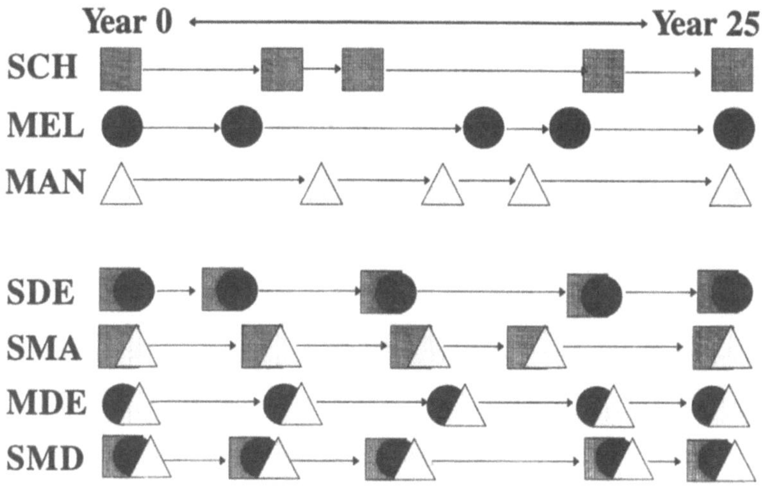

Fig. 5. Longitudinality: hypothetical monomorphous course. See Fig. 4 for abbreviations

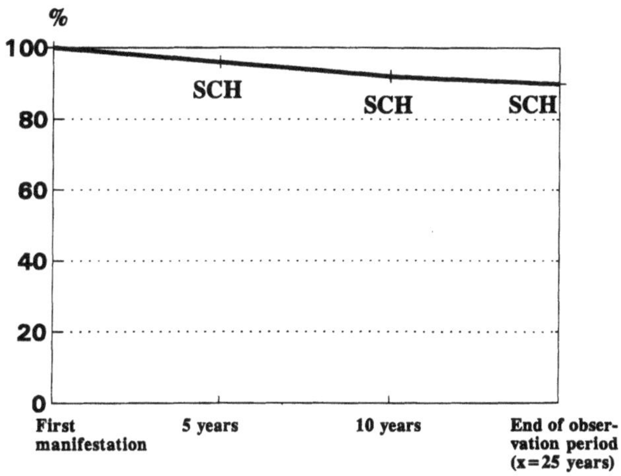

Fig. 6. Longitudinal stability of course of schizophrenia (*SCH*)

of episode, for example, the paranoid type, changes into another one, for example, into the catatonic type, or positive into negative etc., as Figs. 7, 8 show.

We can peak of relatively good "external" longitudinal stability in schizophrenics but of very weak "internal" stability (Fig. 7b; Deister and Marneros 1993; Marneros et al. 1991b).

Of the patients beginning with a *melancholic* episode in 21% this changed during the course into another episode (Fig. 9). Interestingly a melancholic episode does not only change into a manic episode but also into a schizophrenic, schizodepressive, or schizomanic episode (Marneros et al. 1991c).

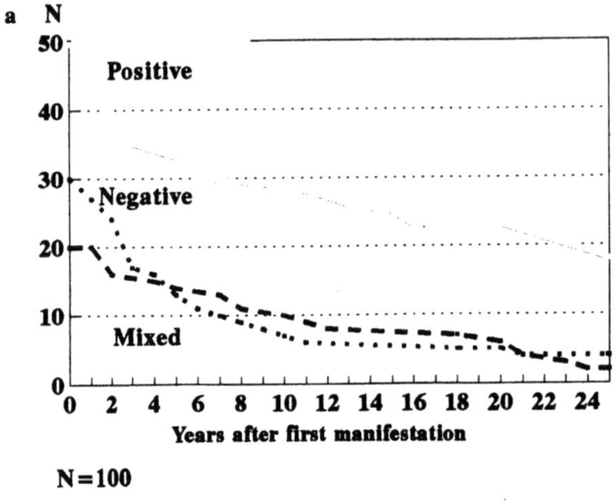

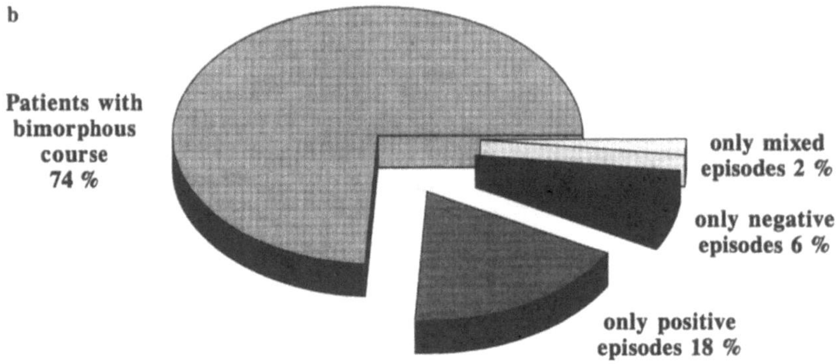

Fig. 7a,b. Internal longitudinal stability of schizophrenia

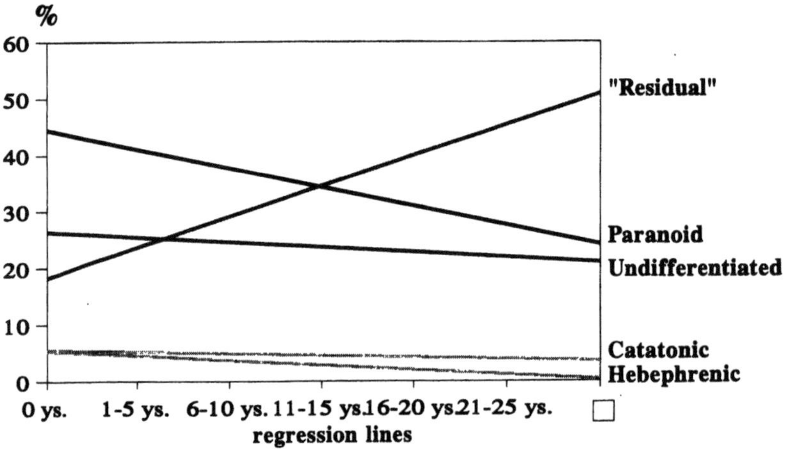

Fig. 8. ICD-10: frequency of various schizophrenic subtypes in a 5-year period

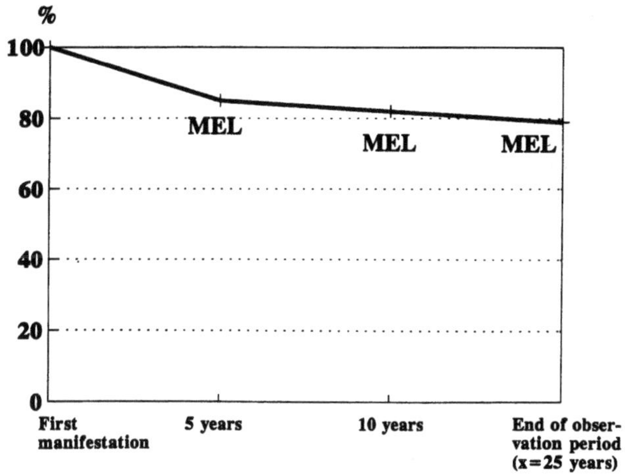

Fig. 9. Longitudinal stability of course of melancholic (*MEL*) episode

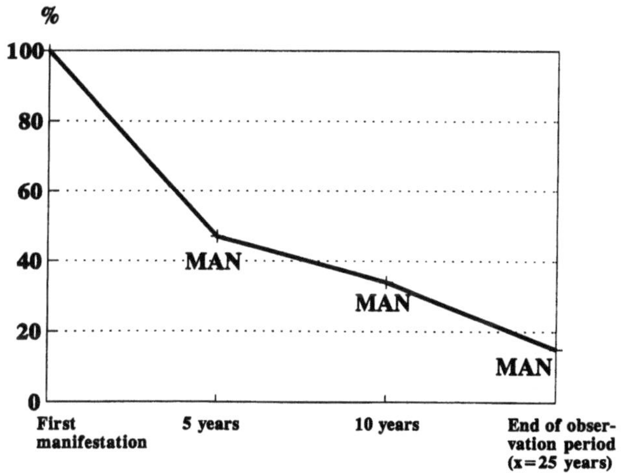

Fig. 10. Longitudinal stability of course of manic (*MAN*) episode

Manic initial episodes almost always changed! Only in three individuals were no changes found during the longitudinal course (Fig. 10). The change occurred not only within the affective group, i.e., not only into melancholia, but also (in 38% of the patients) into schizophrenic or schizoaffective episodes (Marneros et al. 1991c).

Of the *schizodepressive* initial episodes of the Cologne study 42% changed into another type of episode (Fig. 11) (Marneros et al. 1991c). All of the manic-depressive mixed episodes (Fig. 12) and the great majority of schizomanic-depressive episodes (Fig. 13) changed into another type over a long-term course.

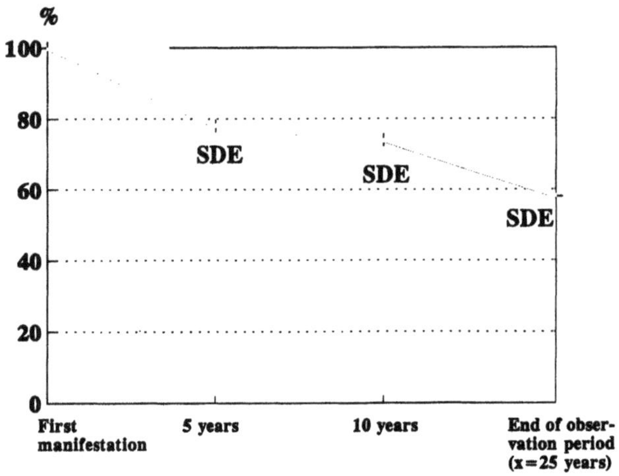

Fig. 11. Longitudinal stability of course of schizodepressive-mixed (*SDE*) episode

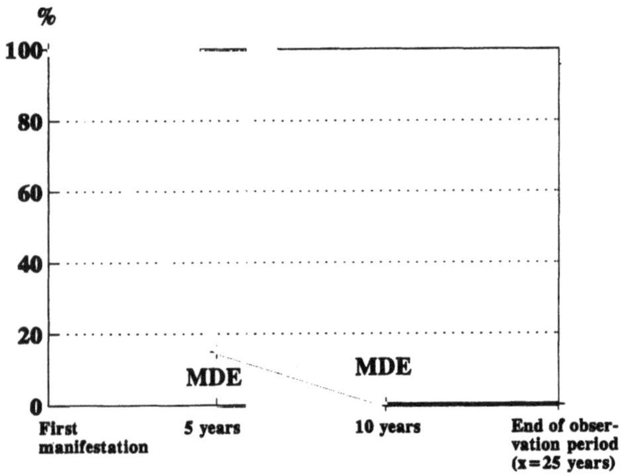

Fig. 12. Longitudinal stability of course of manic-depressive-mixed (*MDE*) episode

If we consider the phenomenological aspect, then we realize that schizophrenic initial episodes (although an "external" stability exists) and the depressive types of episodes (that means melancholic and schizodepressive episodes) are the most stable and it seems that, in contrast, the manic factor alone or in combination with other symptoms is a reason for instability.

Theoretical Considerations

As we reported elsewhere (Marneros et al. 1991c) every syndrome can change longitudinally into every other syndrome, as is demonstrated in Fig. 14. How

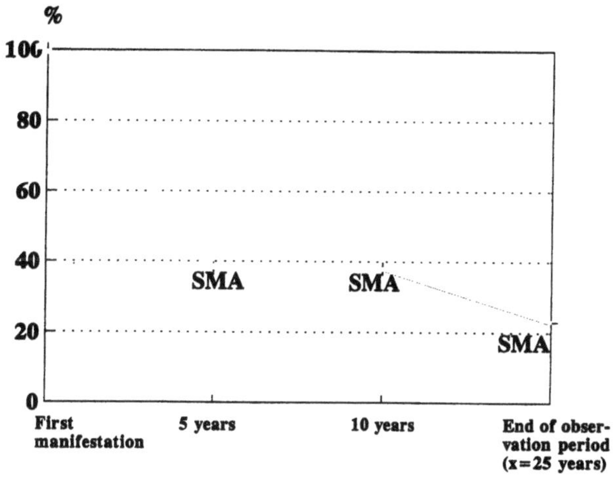

Fig. 13. Longitudinal stability of course of schizomanic (*SMA*) episode

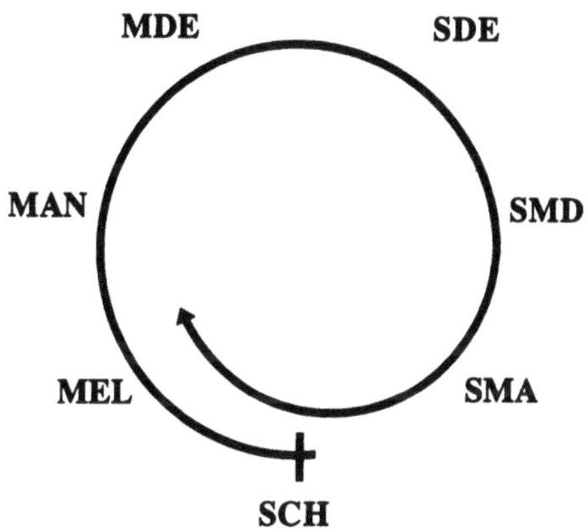

Fig. 14. Longitudinal change of one syndrome into any other (see Fig. 4 for abbreviations)

can we explain the polymorphism of the longitudinal course and the change from one syndrome into another? There are at least three possibilities: Firstly, we can speak of a "one-shot theory," meaning one syndrome, one etiology. In every new manifestation we would have a different disease. Perhaps it is so, but it is difficult to believe.

The second possbility is that we assume a "universal-genesis" of psychoses as Rennert (1965) did. That means that *all types* of syndromes can be manifested within one single course; it also means there are many syndromes but only one etiology. This is the basis of unitary psychosis or of the "universal genesis" of psychoses (Rennert 1965).

The third possibility is that we speak of a "bridge theory": Courses with syndrome change (polymorphous course) are bridges between distinct diseases (having monomorphous course), and so a continuum of various diseases exists.

Outcome

The third important element of Kraepelin's bimodal system is the outcome. If the bimodal distribution of outcome parameters is true, then we have to expect the following distribution, according to Fig. 15: Worse outcome of schizophrenia, in general, and good outcome of affective disorders (some affective patients could have more or less worse outcome). But that is true only if we ignore the schizoaffective disorders. If we do not, then we will find that the distribution of outcome parameters regarding both frequency and severity of persisting alterations change (Figs. 16, 17; Marneros et al. 1992a,b).

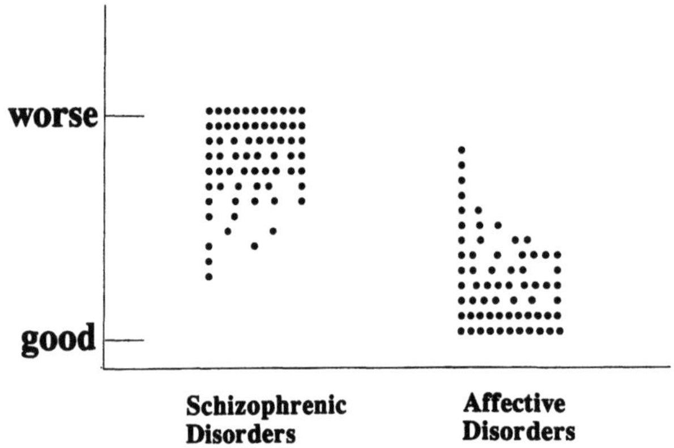

Fig. 15. Hypothetical outcome

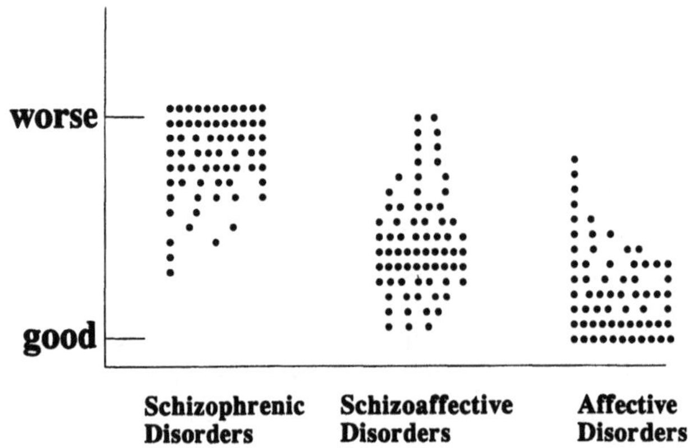

Fig. 16. Clinical outcome

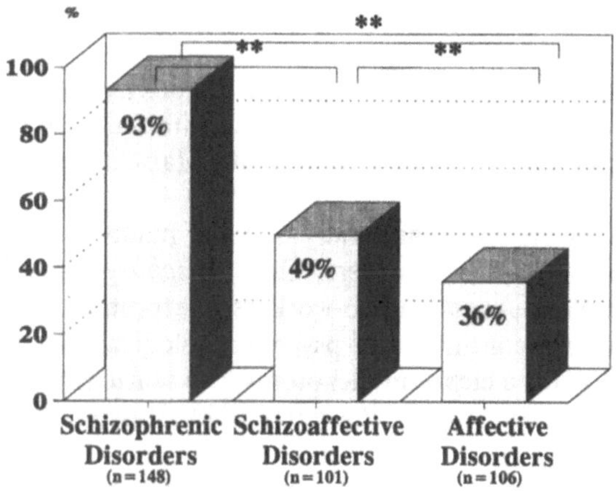

Fig. 17. Persisting alterations

Table 1. Phenomenology of persisting alterations in schizophrenic, schizoaffective, and affective disorders

	Schizophrenic disorders	Schizoaffective disorders	Affective disorders
Depletion syndrome	+	–	–
Apathetic-paranoid syndrome (resp. apathetic-hallucinatory syndrome)	+	+	–
Adynamic deficiency syndrome	+	+	–
Chronic psychosis	+	–	–
Structural deformation	+	+	–
Slight asthenic insufficiency syndrome	+	+	+
Chronic subdepressive syndrome	–	+	+
Chronic hyperthymic syndrome	–	+	+

The phenomenology of persisting alterations of "endogenous" psychoses after a long-term course is also interesting. We allocated the persisting alterations in eight clusters (Marneros et al. 1991a,b). If we compare only the phenomenology of persisting alterations in schizophrenic disorders and affective disorders we will find a nearly total separation. We have only one common syndrome, namely, the "slight asthenic insufficiency syndrome" (Table 1) which, however, is very unspecific, and very ubiquitous. But if we also consider the schizoaffective disorders (Table 1), then we have a bridge between schizophrenic and affective disorders, also in regard of the phenomenology of persisting alterations (Marneros et al. 1991a,b).

Neither frequency nor severity nor phenomenology of persisting alterations in endogenous disorders has a bimodal distribution.

Conclusions

The following can be concluded: The findings of the great majority of longitudinal work concerning schizoaffective disorders is compatible with the findings of our studies (overviews in Marneros and Tsuang 1986, 1990; Marneros 1989; Marneros et al. 1991a–c, 1992a,b).

In particular Angst, already in his important study of 1966, made note of the "atypical" position of schizoaffective disorders, the "*Mischpsychosen*" (Angst 1966). In his subsequent work Angst and co-workers interpreted these and similar findings in the sense of a continuum of psychopathological groups and in the sense of the impossibility for a clear-cut dichotomy into schizophrenia and affective disorders (Angst 1983; Angst and Scharfetter 1990; Stassen et al. 1988).

The above findings opposed the views of Cloninger et al. (1985) based on a prospective follow-up family study of 500 psychiatric outpatients. They found that four symptoms tend to occur together as a stable syndrome throughout the 6- to 12-year follow-up: Persecutory delusions, delusions of control, firmly fixed mood-incongruent delusions, und auditory hallucinations.

The findings presented support the criticism of Crow (1991) who qualified the Cloninger study as "indecisive." Additionally, none of the European long-term studies on schizophrenia found a symptomatological stability (Bleuler 1972; Ciompi and Müller 1976; Huber et al. 1979; Marneros et al. 1991b). The symptomatological changeability, not the stability, characterized the long-term course of psychotic disorders (Janzarik 1968). Additional theoretical and empirical psychopathological arguments such as that of Kendell and Brockington (1980) are discussed in Mundt's paper (1992). The family studies supporting the longitudinal findings (Maier 1992; Crow 1991; Angst and Scharfetter 1990) are also reported in this volume (see the chapter by Tsuang et al.).

The above findings make a clear dichotomy into two diseases, namely, schizophrenia and affective disorders, or even into three diseases, for instance, with the entity "schizoaffective disease," highly questionable.

The "one-shot theory" cannot be supported, but a "universal genesis" of psychotic disorders in the sense of one etiology for all psychotic syndromes cannot be proved either. Nevertheless, the findings of the Cologne study demonstrated above support the assumption of the existence of significant links among psychopathologically defined groups (some of these subgroups were shown here but certainly one can find many more subgroups). The longitudinality of mental disorders shows that psychopathologically defined groups form a bridge between longitudinally monomorphous diseases, building in this way a continuum.

References

Angst (1966) Zur Ätiologie und Nosologie endogener depressiver Psychosen. Eine genetische, soziologische und klinische Studie. Springer, Berlin Heidelberg New York
Angst J (1978) The course of affective disorders. II. Typology of bipolar manic-depressive illness. Arch Psychiatr Nervenkr 226:65–73
Angst J, Scharfetter C (1990) Schizoaffektive Psychosen – ein nosologische Ärgernis. In: Lungershausen E, Kaschka WP, Witkowski RJ (eds) Affektive Psychosen. Schattauer, Stuttgart
Baillarger M (1854) De la folie a double forme. Ann Med Psychol 6:369–389
Berrios GE, Beer D (1992) Unitary psychosis in English speaking psychiatry: a conceptual history, parts 1 and 2. Thieme, Stuttgart (History of psychiatry)
Bleuler M (1972) Die schizophrenen Geistesstörungen im Lichte langjähriger Kranken- und Familiengeschichten. Thieme, Stuttgart
Ciompi L, Müller C (1976) Lebensweg und Alter der Schizophrenen. Eine katamnestische Langzeitstudie bis ins Senium. Springer, Berlin Heidelberg New York
Cloninger CR, Martin RI, Gure SB, Clayton PJ (1985) Diagnosis and prognosis in schizophrenia. Arch Gen Psychiatry 42:15–25
Crow TJ (1991) The demise of the Kraepelinian binary concept and the etiological unity of psychoses. In: Marneros A, Andreasen NC, Tsuang MT (eds) Positive versus negative schizophrenia. Springer, Berlin Heidelberg New York
Deister A, Marneros A (1993) Subtypes in schizophrenic disorders: frequencies in long-term course and premorbid features. Soc Psychiatry Psychiatr Epidemiol 28:164–171
Fallret JP (1851) De la folie circulaire ou forme de maladie mentale characterisée par l'alternative régulière de la manie et de la mélancolie. Bull Acad Natl Med (Paris) (cited in Angst 1987)
Huber G, Gross G, Schüttler R (1979) Schizophrenie. Eine verlaufs- und sozialpsychiatrische Langzeitstudie. Springer, Berlin Heidelberg New York
Janzarik W (1968) Schizophrene Verläufe. Eine strukturdynamische Interpretation. Springer, Berlin Heidelberg New York
Kendell RE, Brockington IF (1980) The identification of disease entities and the relationship between schizophrenic and affective psychoses. Br J Psychiatry 137
Kleist K (1928) Über zykloide, paranoide und epileptoide Psychosen und über die Frage der Degenerationspsychosen. Schweiz Arch Neurol Psychiatr 23:3–37
Kleist K (1953) Die Gliederung der neuropsychischen Erkrankungen. Monatsschr Psychiatr Neurol 125:526–554
Kraepelin E (1899) Psychiatrie, 6th edn. Barth, Leipzig
Kraepelin E (1920) Die Erscheinungsformen des Irreseins. Z Gesamt Neurol Psychiatrie 62:1–29
Leonhard K (1957) Aufteilung der endogenen Psychosen. Akademie Verlag, Berlin
Leonhard K (1980) Aufteilung der endogenen Psychosen, 5th edn. Akadamie, Berlin
Maier W (1992) Kontinuität und Diskontinuität funktioneller Psychosyndrome im Lichte von Familienstudien. Thieme, Stuttgart
Marneros A, Tsuang MT (eds) (1986) Schizoaffective disorders: present level and future perspectives. In: Marneros A, Tsuang MT (eds) Schizoaffective psychoses. Springer, Berlin Heidelberg New York
Marneros A, Deister A, Rohde A (1988) Syndrome shift in the long-term course of schizoaffective disorders. Eur Arch Psychiatr Neurol Sci 238:97–104
Marneros A, Deister A, Rohde A (1989) Unipolar and bipolar schizoaffective disorders: a comparative study. II. Long-term course. Eur Arch Psychiatr Neurol Sci 239:164–170
Marneros A, Tsuang MT (eds) (1990) Affective and schizoaffective disorders. Similarities and differences. Springer, Berlin Heidelberg New York

Marneros A, Andreasen NC, Tsuang MT (eds) (1990) Negative versus positive schizophrenia. Springer, Berlin Heidelberg New York
Marneros A, Deister A, Rohde A (1991a) Affektive, schizoaffektive und schizophrene Psychosen. Eine vergleichende Langzeitstudie. Springer, Berlin Heidelberg New York
Marneros A, Deister A, Rohde A (1991b) Long-term monomorphism of negative and positive schizophrenic episodes. In: Marneros A, Andreasen NC, Tsuang MT (eds) Negative versus positive schizophrenia. Springer, Berlin Heidelberg New York
Marneros A, Deister A, Rohde A (1991c) Stability of diagnosis in affective, schizoaffective and schizophrenic disorders. Cross-sectional versus longitudinal diagnosis. Eur Arch Psychiatr Neurosci 241:187–192
Marneros A, Deister A, Rohde A (1992a) Comparison of long-term outcome of schizophrenic, affective and schizoaffective disorders. Br J Psychiatry 161 [Suppl 18]:44–51
Marneros A, Deister A, Rohde A (1992b) Phenomenological constellations of persisting alterations in functional disorders: a comparison. In: Ferrero FP, Haynal AE, Sartorius H (eds) Schizophrenia and affective psychoses. Nosology in Contemporary Psychiatry. Libbey, Rome
Mundt C (1992) Einheit oder Vielfalt von Endogenität der Psychosen. In: Mundt Ch, Saß H (eds) Für und wider die Einheitspsychose. Thieme, Stuttgart
Perris C (1966) A study of bipolar (manic-depressive) and unipolar recurrent depressive psychoses. Acta Psychiatr Scand [Suppl] 194:1–89
Perris C, Brockington IF (1981) Cycloid psychoses and their relation to the major psychoses. In: Perris C et al (eds) Biological psychiatry 1981. Elsevier, Amsterdam
Rennert (1965) Die Universalgenese der endogenen Psychosen. ein Beitrag zum Problem "Einheitspsychose". Fortschr Neurol Psychiatr 33:251–272
Schmidt-Degenhardt M (1992) Einheitspsychose – Begriff und Idee. Thieme, Stuttgart
Stassen HH, Scharfetter C, Winokur G, Angst J (1988) Familial syndrome patterns in schizophrenia, schizoaffective disorders, mania, and depression. Eur Arch Psychiatr Neurol Sci 237:115–123
Vliegen J (1980) Die Einheitspsychose. Geschichte und Problem. Enke, Stuttgart
Wernicke C (1900) Grundriss der Psychiatrie. Thieme, Leipzig

Psychotic Continuum or Distinct Entities: Perspective from Psychopharmacology

H. Y. Meltzer

Introduction

The issue inherent in the title of this workshop, whether schizophrenia is comprised of distinct entities or part of a psychotic continuum, contains within it at least two questions which should be considered separately if they are to be satisfactorily addressed. These are: (a) whether the patients who meet such relatively narrow criteria for schizophrenia as those of the DSM-III-R belong to a larger psychotic continuum, and (b) whether there is more than one distinct entity, i.e., disorder, with distinct etiology and pathophysiology, included within DSM-III-R schizophrenia. I specify DSM-III-R schizophrenia, because there are over a dozen criteria for schizophrenia. Very broad criteria such as those of Kurt Schneider would tend to have subtypes, whereas narrow criteria such as DSM-III-R would tend to be more homogeneous. Moreover, DSM-III-R and the closely related ICD-9/10 criteria are the most widely used criteria for research as well as clinical purposes.

The first issue embedded within the organization of this symposia is whether DSM-III-R schizophrenia, in its totality, is part of a continuum which may include other psychiatric disorders, e.g., manic-depressive disorder, unipolar depression, psychotic subtype, and perhaps organic brain disorders such as temporal lobe epilepsy. This issue may be reformulated as to whether there is a unitary (*einheit liche*) psychosis, of which schizophrenia and psychotic mood disorders, temporal lobe epilepsy, etc., are integral components or, at the least, whether the different forms share significant biological features (Meltzer 1973, 1976). The second issue is whether there are different and distinct schizophrenias (Meltzer 1979), e.g., whether various subtypes of schizophrenia, such as the positive and negative subtypes, paranoid and disorganized subtype, type I and type II, etc., represent unique entities. This latter issue is to be distinguished from the issue of syndromes within schizophrenia, e.g., positive, negative, and disorganized, as discussed by Peter Liddle (1995). Syndromes are conceptualized as dimensional processes which may be part of a single disease entity called schizophrenia (Crow et al. 1986). These two issues have been debated for nearly a century without firm agreement.

Department of Psychiatry, University Hospitals of Cleveland, Hanna Pavilion, Room B-68, 2040 Abington Road, Cleveland, OH 44106-5000, USA

This article will consider the role of psychopharmacology in determining whether schizophrenia is one component of a unitary psychosis or whether there are subtypes of schizophrenia which can be defined, in part, by differences in response to pharmacological challenges or neuroleptic treatment.

Before doing so, I should like to review the results of biological studies of the neuromuscular system in the psychoses, which provide the background for my perspective on these issues. In a series of studies of patients who met criteria for either schizophrenia, bipolar disorder, or unipolar depression it was found that schizophrenic patients, by definition psychotic at some phase of their illness and psychotic unipolar and bipolar (manic or depressed phase) patients could be distinguished from nonpsychotic depressed patients and normal controls by a variety of neuromuscular measures, e.g., (a) increased serum creatine kinase (CK) activity of the skeletal muscle type; (b) the number of atrophic muscle fibers and other types of skeletal muscle fibers suggestive of neurogenic muscle disease. Some of these abnormal muscle fibers were likely to be neurodevelopmental in nature because of their presence in hereditary disorders, e.g., nemaline rod body-containing fibers, central core disease, while others were more likely to be neurotoxic sequelae, e.g., fiber-type grouping); (c) ultrastructural changes in muscle fibers such as Z-band streaming; (d) increased sprouting and branching of subterminal motor nerves (suggestive of a neurodegenerative process); and (e) abnormalities of motor nerve physiology, such as diminished H-reflex recovery curve (Meltzer 1973, 1975, 1976, 1979; see Meltzer and Crayton 1975 for review). Increased motor unit territories were also found in schizophrenic patients (Crayton et al. 1977), consistent with the evidence of neurodegeneration and regeneration obtained from the study of skeletal muscle and subterminal motor nerve pathomorphology. It was suggested, on the basis of the presumptive neuronal loss leading to sprouting and branching of subterminal motor nerves, that "excess neuronal branching might occur to a significant extent in the central nervous system of an individual with schizophrenic or affective psychoses. The extent of disability might depend on what type of new connections were made. It seems improbable that there should be denervation and reinnervation in the peripheral motor nerves of psychotic patients without comparable abnormalities or attempts at restoration of function in the central nervous system" (Meltzer 1976). The evidence from this group of studies was supportive of the hypothesis that the biology of schizophrenia was part of a psychotic continuum which included psychotic mood disorders. No evidence for differences between subtypes of schizophrenia were found in this series of studies.

These data suggested, from a biological point of view, that schizophrenia and the psychotic mood disorders share a common pathophysiology, but not that they were the same disease entity. These conclusions are consistent with the continuum hypothesis of Crow (1990). Evidence for similar abnormalities in ventricular brain ratio, prefrontal cortical sulcal widening, developmental abnormalities in entorhinal cortex, and smooth-pursuit eye movement deficits in both schizophrenia and mood disorders have been reported in recent years. It is beyond the scope of this article to review these common biological abnormalities, which may be discussed, in part, by Van Kammen elsewhere in this volume. However, they are consistent with and provide the basis for further evaluation of the continuum hypothesis.

The Psychopharmacologic Perspective: An Orientation

As previously mentioned, there are at least two ways in which psychopharmacology can be used to address the issue of whether schizophrenia is comprised of distinct entities or is part of a psychotic continuum. These are: (a) comparing responses to single doses of drugs given as "challenge" agents, and (b) clinical, behavioral, and biological responses to chronic drug treatment. The former approach is actually a blend of psychopharmacology and biological psychiatry, integrating the known mechanism of action of drug with theories about the etiology or pathophysiology of psychiatric disorders, in order to determine if a specific neurotransmitter system is abnormal in a specific disease or perhaps its subtypes. The latter approach involves the classic attempt to identify subtypes of a disorder by the presence or absence of a therapeutic response to an agent which acts on the core disease process (Kendall 1975). Studies which compare drug effects in both mood disorders and schizophrenia are relevant to the continuum hypothesis. Variability in antipsychotic drug response within patients who meet criteria for schizophrenia is relevant to the question of whether there are subtypes of schizophrenia. The evidence that neuroleptic-resistant and -responsive schizophrenia are separate subtypes will be briefly reviewed.

Some aspects of the general issue of using treatment-response data to subtype schizophrenia have been discussed by Csernansky et al. (1985) and Brown and Herz (1988). Csernansky et al. (1985) discussed some of the issues that may affect neuroleptic response: symptom profile, prior course of illness, brain atrophy, the possible basis for variability in response over time, e.g., alterations in DA receptor sensitivity, dose and plasma concentrations. Brown and Herz (1988) identified several conditions which they propose must be satisfied before the variability of response to a treatment may be utilized for nosologic purposes and be pertinent to the pathophysiology of an illness: (a) ability to characterize patients as responders and nonresponders; (b) consistency of responses during recurrences; (c) relation of response to "physiologically meaningful" variables such as symptoms, family history, biological changes, etc.; and (d) a direct effect of the treatment on the "fundamental pathophysiology" of the disorder.

Neither of the two above-mentioned papers definitively assessed the issue of whether the response to neuroleptic drugs in schizophrenia can be helpful to subtype schizophrenia, but they did provide useful suggestions as to the process to be followed. Among many studies attempting this are those of Crow (1980), Overall and Rhodes (1982), Kolakowska et al. (1985), Garver et al. (1988), Schulz et al. (1988), Bartkó et al. (1989), Whelton et al. (1992), and Sautter et al. (1993). None of these studies meet the criteria suggested by Brown and

Herz (1988), usually because of failure to study patients over multiple episodes and to show relationships between response or nonresponse to the fundamental pathophysiology of the disorder.

Acute Administration of Psychopharmacologic Agents

Single doses of agents which act as direct or indirect agonists at specific neurotransmitter receptors can be used to assess the status of neurotransmitter systems, which are believed to be implicated in the etiology and pathophysiology of schizophrenia. For example, it has long been suspected that the three neurotransmitters dopamine (DA), serotonin (5-HT), and norepinephrine (NE) are involved in the etiology or pathophysiology of schizophrenia (Meltzer and Stahl 1976; Meltzer 1987). The possibility that subtypes of schizophrenia are characterized by specific abnormalities of these (and other) neurotransmitters can be studied by monitoring the acute effects of specific direct or indirect agonists on a variety of measures, e.g., plasma hormones, neurochemicals, brain blood flow, behavior, and physiological measures. Some of this literature will now be reviewed.

Apomorphine

The prototype of the pharmacologic deconstruction (i.e., identification of component parts) of schizophrenia is a series of studies using apomorphine, a direct-acting D_1/D_2 DA agonist. At the dose range used (0.05–0.10 mg/kg), apomorphine is expected to have mainly postsynaptic DA receptor effects, e.g., behavioral activation or hormone responses, but it cannot be ruled out that it might also produce presynaptic DA autoreceptor stimulation (Meltzer 1982). There are many types of responses to an apomorphine challenge, e.g., endocrine, neurochemical, behavioral, somatic, which are potentially useful to determine whether schizophrenia is composed of more than one unique subtype and whether it may be differentiated from other forms of mental illness, e.g., depression and manias, as well as from normal controls (Meltzer et al. 1981). PRL and especially the GH response to apomorphine have been most studied. The results suggest that diminished GH response is more likely in schizophrenia than in mood disorders (Meltzer et al. 1981, 1982, 1984a; Hirschowitz et al. 1986). We have observed that the magnitude of this response is inversely related to the duration of illness (Meltzer et al. 1981, 1982, 1984a). These results have been replicated by Zemlan et al. (1986a). Brown et al. (1988) also found a decline in the apomorphine-induced GH response with duration of illness in 5/12 patients, while the response increased in three others. However, factors other than duration of illness also influence the apomorphine-induced GH response and are possibly related to the concept of discrete subtypes of schizophrenia: poor premorbid function (Malas et al. 1987); schizophrenia-spectrum psychopathology in first-degree relatives

(Sautter et al. 1987); negative symptoms (Ferrier et al. 1984; Crow et al. 1986); Schneiderian symptoms (Whalley et al. 1984); psychotic symptom clusters (Zemlan et al. 1986b); and thought disorder (Zemlan et al. 1986a), but not tardive dyskinesia (Muller-Spahn et al. 1984). The effect of apomorphine on regional cerebral blood flow in the prefrontal cortex (Daniel et al. 1989) and striatum (Cleghorn et al. 1991) has also been studied.

All of these studies, because of their relevance to the issue of DA receptor sensitivity in schizophrenia and the role of DA in influencing specific symptoms and aspects of the pathophysiology of schizophrenia, provide information about subtypes versus continuum. However, the multiplicity of factors which affect the magnitude of the GH response, blood flow, etc., including pharmacokinetic factors, residual effects of prior neuroleptic treatment, and the status of other neurotransmitter systems which influence specific components of response make if difficult to draw firm conclusions about subtypes of response. The issue of how fundamental is the quantitative variation among schizophrenics in the various responses to apomorphine is still under investigation. It may be that responsivity to apomorphine reflects neurochemical differences that determine catalepsy and influences symptomatic response to neuroleptic treatment based upon fundamental differences in pathophysiology (Hirschowitz et al. 1982).

Methylphenidate, Amphetamine and L-DOPA

A considerable effort to pharmacologically deconstruct schizophrenia based upon the DA hypothesis of its etiology has also employed indirect DA agonists, e.g., methylphenidate, amphetamine, and L-DOPA. Response to each of these agents is sensitive to one or more aspects of the dopaminergic system, e.g., presynaptic effects such as uptake of L-DOPA and its decarboxylation to DA, release of newly synthesized DA, and reuptake of DA, as well as a variety of postsynaptic effects. It is conceivable that differences in these or other aspects of DA metabolism might be clinically relevant and could be revealed by challenge studies.

Van Kammen and colleagues have indicated that differences in the behavioral response to i.v. *d*-amphetamine may predict response to lithium (Van Kammen et al. 1981) or pimozide (Van Kammen et al. 1982a,b). In a follow-up study of 30 schizophrenic patients, Van Kammen and Bornow (1988) noted that i.v. amphetamine improved both negative and depressive symptoms, and that the amphetamine-induced decrease in negative symptoms predicted improvement on negative symptoms while on pimozide. They suggested these results did not support the hypothesis of Crow (1980) that negative symptoms were irreversible and were unrelated to the dopaminergic system. Other investigators have reported that chronic administration of amphetamine to neuroleptic-treated schizophrenic patients can improve negative symptoms or the defect state (Angrist et al. 1982; Cesarec and Nyman 1985). The latter study is of particular interest. These authors reported that a positive response to amphetamine was correlated with the presence of florid symptomatology, while subjects with negative symptoms seemed totally unresponsive. This suggests that the amphetamine response may be related to important neurochemical differences relevant to psychopathology, although one can comfortably reject the hypothesis that positive and negative symptoms segregate with subtypes of schizophrenia. One may still postulate that patients without positive or negative symptoms have no abnormalities in specific areas of dopaminergic function, e.g., subcortical and cortical, respectively, while still sharing a core biological deficit that might not involve DA directly, e.g., one which affected the third cluster of psychotic symptoms, disorganization. Another study suggestive of the possibility that the response to *d*-amphetamine may identify a subtype of schizophrenia is that of Pandurangi et al. (1989). These investigators found that worsening of positive psychotic symptoms during an amphetamine challenge was more likely in schizophrenic patients without ventricular enlargement, and that these patients went on to have a better response to neuroleptic drugs. They noted that these results are, in general, consistent with the hypothesis of type I/II schizophrenia (Crow 1980).

One study of the effect of amphetamine on regional cerebral blood flow (RCBF) in schizophrenia is consistent with an abnormal dopaminergic system in some schizophrenics (Wolkin et al. 1987), but another showed no differences in the effect of amphetamine on RCBF between schizophrenics and normal controls. The study by Wolkin et al. (1987) indicated that cortical atrophy was associated with a blunted metabolic response to amphetamine. In a further elaboration of this approach, the effect of amphetamine on the performance of the Wisconsin Card Sort test (WCST) and on dorsolateral cortical blood flood was compared in neuroleptic-treated schizophrenic patients and normal controls. In this case, six patients showed general improvement, but these patients had large ventricles. No patients worsened (Goldberg et al. 1991). Amphetamine enabled the schizophrenic patients to activate their dorsolateral prefrontal cortex, and this was correlated with performance on the WCST task (Daniel et al. 1991).

Similar studies have been carried out with methylphenidate. There appears to be a subgroup of schizophrenic patients who have a positive response (exacerbation) of positive and negative symptoms to a methylphenidate infusion. This was associated with rapid relapse after neuroleptic withdrawal (Lieberman et al. 1984, 1987). Methylphenidate-induced exacerbation occurred in 23/38 (61%) first-break schizophrenics. This was not related to neuroleptic response, but was associated with increased vulnerability to EPS (Jody et al. 1990). No evidence of a subgroup of schizophrenic patients who were specifically vulnerable to methylphenidate-induced activation of psychotic symptoms or biological responsivity (GH response, plasma HVA) was noted in another study (Sharma et al. 1990, 1991).

In summary, administration of direct or indirect DA agonists to neuroleptic-treated or drug-free schizophrenic patients has produced a complex pattern of results, which suggests that there may be subgroups of schizophrenic patients who are more or less responsive to these agents as determined by immediate behavioral activation, changes in biological indices, and response to neuroleptic administration or withdrawal. These responses appear stable in some patients but unstable in others. There has been too little examination of this issue to draw conclusions about state-versus-trait characteristics. There has been no comparison of direct-acting apomorphine versus indirect-acting drugs which might permit clarification of their interrelationships. Presumably, if the apomorphine responses are postsynaptic, apomorphine should cause an activation of psychopathology similar to that of *d*-amphetamine or methylphenidate, but it does not. Because of these uncertainties, it would appear that no conclusion can be drawn on the basis of this work as to whether schizophrenia is a subtype. The results tend to indicate a dopaminergic abnormality in some patients with schizophrenia, but the significance of this is unknown. They also indicate that DA is relevant to positive *and* negative symptoms, but it is unlikely that this is specific to schizophrenia.

Neuroleptics

Another method of assessing the dopaminergic system via hormone responsivity is to administer DA antagonists rather than direct or indirect DA agonists and to measure the increase in plasma PRL which results from D_2 receptor blockade (Meltzer and Fang 1976). This increase is, in fact, the net result of complex pharmacokinetic and pharmacodynamic factors. The latter include not only the responsivity of D_2 receptors on the lactotrophes and the available stores of PRL, but also turnover of PRL in plasma, rate of DA release from the tuberoinfundibular neurons, transport of DA from the median eminence to the pituitary, etc. Two studies using low doses of neuroleptic drugs showed no differences between schizophrenic patients and controls (Gruen et al. 1978; Meltzer et al. 1981). Blunted PRL responses to haloperidol have been reported in three studies (Wolf et al. 1982; Asnis and Ryan 1983; Keks et al. 1990). Keks et al. (1992) reanalyzed their earlier data, using multiple systems of diagnosis, and reported that schizophrenic patients diagnosed according to Kraepelinian criteria were separable from the non-Kraepelinian patients who still met criteria for schizophrenia by various other systems. Studies such as this can be only suggestive of genuine subtypes because so many factors in the PRL response were uncontrolled. However, there is considerable other evidence that Kraepelinian schizophrenia may represent a distinct subtype (Keefe et al. 1990).

Serotonergic Challenges and Subtyping Schizophrenia

There is considerable interest in the possibility of a serotonergic component to the etiology of schizophrenia (Bleich et al. 1988) and a serotonergic-dopaminergic component (Meltzer 1989a). Various serotonergic drugs have been used as challenge agents to subtype schizophrenia, including MK-212, m-chlorophenylpiperazine (mCPP), fenfluramine, and ipsapirone. Cimetidine, a drug usually not considered to act via a serotonergic mechanism, nevertheless stimulates PRL secretion by a serotonergic mechanism. This response is markedly blunted in male schizophrenia patients compared with normal controls but is normal in female schizophrenic patients (Meltzer et al. 1993), suggesting a pharmacologic means of subtyping schizophrenia by gender. We have also found that the temperature response to MK-212, a direct-acting 5-$HT_{2A/2C}$ agonist, is consistently hyperthermic in normal males but is diminished in male schizophrenic patients, with some patients actually having a hypothermic rather than a hyperthermic response (Lee et al. 1991). There is intriguing evidence of abnormal behavioral and endocrine responses to mCPP (Iqbal et al. 1991; Kahn et al. 1992) and fenfluramine (Lerer et al. 1988). Larger samples must be studied to determine if patients with abnormal responses to these agents differ in genetic load, age or type of onset, phenomenology, cognitive function, biological measures such as ventricular brain ratio, response to treatment, etc.

Serotonergic Challenges and the Continuum Hypothesis

There is extensive evidence for the role of 5-HT in affective disorders (Meltzer and Lowy 1987). Neuroendocrine challenges, as well as depletion of serotonin with parachlorophenylalamine (PCPA) and the tryptophan depletion strategy, have been useful in identifying the serotonergic deficit in depression and mania. The pattern of responses to these interventions is not supportive of a continuum hypothesis. For example, the cortisol responses to 5-HTP (Meltzer et al. 1984b) and ipsapirone (Lesch et al. 1990) are enhanced and blunted in depression, respectively, but both are normal in schizophrenia. Tryptophan depletion leads to a reversal of antidepressant response in major depression (Delgado et al. 1990) but has little effect in schizophrenia.

Pharmacologic Treatment and Nosology

Among the early theorists who advocated that differential response to specific neuroleptic drugs may identify specific subtypes of schizophrenia were Hollister et al. (1976). However, the relationships between response to specific neuroleptic drugs and symptom profiles identified by Hollister and his colleagues were often not reproducible, leading to abandonment of this approach (see Csernansky et al. 1985 for review). The remission of psychotic symptoms in schizophrenic patients in the absence of neuroleptic drug treatment was suggested to define subtypes of schizophrenia by Goldstein (1970) and Carpenter et al. (1977), but it is more likely that this represents part of a continuum of response to neuroleptic treatment together with spontaneous remission and placebo response rates (Young and Meltzer 1980). Overall and Rhodes (1982) utilized Brief Psychiatric Rating Scale (BPRS) data from 1438 patients treated with neuroleptic drugs–diagnostically broad criteria for schizophrenia–to identify four phenomenological subtypes and to show that the 6-week response to neuroleptics ranged from a low of 8.5% in the withdrawn-disorganized type to a high of 39.3% in the hostile-suspicious subtype. The two groups showed placebo-response rates of 9.5 and 29.4%, suggesting virtually no drug effect in the withdrawn-disorganized subtype. Because these data came from combining data from multiple studies, the results can be considered only suggestive. The differential response of positive and negative symptoms to neuroleptic drugs was suggested to be one of three factors useful to subtype schizophrenia: neuroleptic response, brain atrophy, and symptom pattern (Crow 1980). However, the concept of integrating structural brain deficits, symptom profile, and neuroleptic response to subtype schizophrenia has not proven valid, in that all three factors often do not segregate as proposed by Crow (1980) (e.g., Meltzer 1985; Losonczy et al. 1986).

The efficacy of somatic treatments provides an inconclusive data base concerning the continuum hypothesis. On the one hand, there is little evidence that some drugs effective in specific mood disorders, such as lithium carbonate or the various classes of antidepressants and mood stabilizers (e.g., carbamazepine or valproate), have any primary efficacy in most patients with schizophrenia. These thymoleptic drugs may be of value when concomitant mood symptoms are superimposed on a schizophrenia picture or when schizoaffective disorder is present (Garver et al. 1988), but most studies suggest they are of no value in treating positive or negative symptoms in schizophrenia (Meltzer 1992). On the other hand, neuroleptic drugs, and especially ECT and clozapine, are effective for treating the primary symptoms of mania, psychotic depression, and schizophrenia (see Meltzer 1992 for references; Sajatovic and Meltzer 1993; Calabrese et al. 1991). This broad spectrum of action suggests the possibility for some overlapping of biological characteristics among these disorders, although it would be incorrect to conclude that the ability of these therapies to partially resolve many, but not all, cases of these disorders means they have the same primary abnormalities. Rather, it is more prudent to conclude that these nonselective therapies can bring about a series of changes in neurotransmission that are useful in a group of disorders that may, in fact, have entirely distinct etiologies. However, these nonselective therapies may indicate that these disorders share abnormalities of the same neurotransmitters, e.g., 5-HT, DA, NE, GABA, etc. which, though different in nature, may usually be alleviated by these agents. In my opinion, these nonselective therapies constitute evidence in behalf of a partial continuum hypothesis, but I would not suggest abandoing the Kraepelinian dichotomy on the basis of this evidence alone.

The fact that the very potent pharmacotherapies developed for schizophrenia and mood disorders are nevertheless ineffective, to a clinically important extent, in about 15–30% of mood disorder and schizophrenic patients constitutes potentially important evidence for subtyping these disorders. Failure to respond to an agent which is effective in the large majority of patients with apparently similar illnesses can be evidence that etiology or pathophysiology is different between the two groups. Recent molecular studies suggest, for example, that the failure of bromocriptine to suppress prolactin secretion in some cases of prolactinomas is due to a variety of mutations affecting the DA receptor recognition site or signal transduction systems (Caccavelli et al. 1992). It is tempting to use this model to explain the failure of neuroleptics to suppress positive symptoms in some schizophrenic patients, i.e., that there may be an intrinsic difference in how D_2 receptor blockade alters dopaminergic transmission in these cases.

The possibility that neuroleptic resistance identifies a specific type of schizophrenia will now be considered.

Resistance to Antipsychotic Drug Treatment: Does It Reveal Discrete Subtypes?

Using differential response to drug treatment as a means of identifying whether a putative clinical entity is homogeneous or heterogeneous has great surface appeal because of the possibility that variability in treatment response must somehow be related to differences in essential etiologic aspects. Furthermore, a classification would be highly useful as a guide to treatment of newly encountered patients, avoiding clinical trials which have only a small chance of succeeding. From a heuristic viewpoint, identification of subgroups of schizophrenics who respond to agents *not* effective in other groups should be of great help in identifying etiologic factors in the group that does respond, and abnormalities must be of lesser importance, if present at all, in the group that does not respond. Response to a trial of neuroleptic drugs at what should be an adequate dose and duration of time, e.g., 10–20 mg haloperidol/day for 4–12 weeks, has been shown to be a continuum of responses rather than any clearly dichotomous outcome, although for various purposes, classifications into discrete response categories can be made. The potential categorization schemes range from the very simple such as responders and nonresponders, or good and poor responders, to much more differentiated classifications, such as that proposed by Brenner et al. (1990) which identified seven outcome levels: clinical remission, partial remission, slight resistance, moderate resistance, severe resistance, refractory, and severely refractory, on the basis of global function, positive and negative symptoms, and social and vocational function. The severely refractory level is described as having no response whatsoever to neuroleptic treatment. It is noteworthy that while even those considered to be in clinical remission by Brenner et al. (1990) must have no psychotic symptomatology and on global estimation be considered normal, not mentally ill, they may still have some mild negative symptoms. This highlights the difficulty of separating some psychiatric symptoms from normal behavior. The other six levels described by Brenner et al. (1990) clearly represent a continuum of partial response. It is possible that the first and seventh patterns of response might represent distinct subtypes. However, it is apparent that in clinical practice it could prove exceedingly difficult to reliably distinguish which response group a particular patient fell into.

It is most important in using a differentiated scale such as that of Brenner et al. (1990) or specific levels of symptomatology such as those employed by myself and colleagues in the U.S. clozapine multicenter trial (Kane et al. 1988) to identify possible distinct subtypes of antipsychotic drug responders, to consider separately *change* in response to treatment rather than the final absolute level on whatever outcome criteria are utilized. Because of differences in initial severity, some patients who show minimal or no change in response to a specific treatment, and for this reason may be considered poor or nonresponders, may yet have a better final assessment on one or more outcome criteria than patients with a high initial severity who showed a clear, clinically relevant, but partial response to a given intervention. The latter individuals, for the purpose of understanding etiology and pathophysiology in relation to the mechanism of action of drugs, should definitely be considered responders with whatever qualification necessary to describe the limited extent of change during the course of treatment. Percentage changes in rating scales such as the BPRS (Overall and Gorham 1962) or the Positive and Negative Symptom Scale (PANNS; Kay et al. 1987, 1988), which are a function of initial level of severity (Thompson et al. 1994), may be misleading as a means of classifying patients on the basis of response. Two patients with initial BPRS total scores of 70 and 45 (1–70 scoring system), respectively, might both show a decrease in BPRS of 10, based on improvement in the same items. The two patients would have 14.3% and 22% decreases in BPRS, respectively. The first would be a "nonresponder" by the criteria of Kane et al. (1988), while the latter would be a responder.

The nature of the types of psychopathology which change during treatment is also obviously a potential source of confusion regarding response criteria. Improvement in positive symptoms alone, negative symptoms alone, or anxiety/depression alone might produce similar changes in total BPRS scores but must be distinguished if there is to be any chance of identifying homogeneous subtypes or clarifying relationships along a continuum model. For this reason, multiple regression analyses or analyses of variance which adjust for initial level of severity are preferable when analyzing group data to ascertain the relationship between individual pretreatment characteristics and treatment outcome.

From the perspective of an effort to identify subtypes of schizophrenia on the basis of neuroleptic response, one must specify the type(s) of symptomatology which are evaluated, the extent of change, if any, and criteria for initial and final outcome measures. Patients who selectively fail to improve on positive, negative, or disorganization symptoms or some combination of these three syndromes may differ neurobiologically from those with different syndrome response patterns. Such differences would not be apparent if only global measures such as the BPRS total or PANNS total scores were used to categorize clinical response. From the perspective of fundamental biological difference, outcome measures such as social function, "insight", etc. that might be subject to a variety of premorbid, psychosocial and societal influences could prove very difficult to use. Therefore, it seems unlikely that a dichotomous view of neuroleptic resistance will be successful. There may be a variety of separate subtypes of neuroleptic resistance which must be identified before meaningful neurobiological differences between them and neuroleptic responders can be identified. The identification of these subtypes will be possible only with large samples and with adequately reliable and differentiated rating scales for relevant outcome measures.

As pointed out by Strauss and Carpenter (1972, 1978), with regard to typical neuroleptic therapy, and by the author with regard to clozapine (Meltzer 1992), response of schizophrenic patients is best viewed from a multidimensional perspective, with psychopathology, social function, work function, cognition, quality of life, and rehospitalization, among other outcome measures, often showing a surprising degree of independence. Furthermore, core outcome measures such as psychopathology and cognitive function can themselves be further subdivided into relatively independent and clinically significant factors or domains, e.g., positive, negative, and disorganization symptomatology and executive function, recall, working and semantic memory, and attention. Some patients clearly improve in some but not all types of psychopathology, while others show little or no change (Meltzer et al. 1986; Meltzer 1992). Responsivity to antipsychotic drugs of the classical neuroleptic type would identify distinct subtype(s) of schizophrenia if the ability of these agents to decrease dopaminergic activity by blocking DA receptors and/or suppress the firing rate of DA neurons projecting to mesolimbic, mesocortical, and striatal neurons could be shown to be relevant to improvement in positive, negative, or disorganization symptoms. There is some evidence that this is in fact the case. Studies utilizing plasma prolactin elevations (Meltzer and Fang 1976) or occupancy of striatal DA receptors (Wolkin et al. 1989) would appear to suggest that there are some schizophrenic patients for whom decreasing dopaminergic activity has little or no clinical significance, at least at the time during which little responsivity to neuroleptic drugs is present. Establishing that an adequate trial of a neuroleptic drug has, in fact, taken place, for a given individual as well as for the entire study population, is essential before evaluation of response and potential classification. Evaluation for neuroleptic responsivity includes compliance, assuring the use of medication in an appropriate dosage (e.g., 10–20 mg/day haloperidol, or higher if necessary), measurement of plasma levels, if possible, or at least plasma prolactin to assure compliance, an adequate duration of treatment (at least 6 weeks' treatment with optimal dose at steady state), trials of at least two different neuroleptics of different chemical classes, and, for inpatients, an adequate milieu with a good psychosocial treatment program. Comorbidity, i.e., substance abuse, may also influence neuroleptic response (Bowers et al. 1990; Seibyl et al. 1993; Buckley et al. 1994). The response to typical neuroleptics is impaired by concomitant substance abuse (Bowers et al. 1990; Seibyl et al. 1993), but this does not appear to be the case for clozapine (Buckley et al. 1994).

Because of these considerations, it is evident that there is a possibility of misclassification of some schizophrenic patients as poor responders. Many patients may be judged to be poor responders because only one type of outcome measure (e.g., positive, negative, or disorganization symptoms) was considered or because the trial(s) were inadequate for one or more of the reasons given above. Conversely, some patients may be considered good drug responders who really responded because of spontaneous remissions (Young and Meltzer 1980).

Some of these factors may explain why there have been such conflicting claims, beginning with Crow (1980), that large ventricles and/or cortical sulcal widening predicts response to neuroleptic drugs. There have been over 50 studies relevant to this issue. Careful reviews do not support any such relationship (Raz and Raz 1990; Friedman et al. 1992), even though some studies, using relative extremes of responsivity, sometimes report that responders have less cerebral atrophy than nonresponders. We have measured ventricular brain ratio (VBR), third ventricle width (TVW), and prefrontal sulcal widening (PFSW) in neuroleptic-resistant patients (criteria of Kane et al. 1988) and neuroleptic-resistant schizophrenic patients (Table 1) using CT. No significant differnces were found.

Studies of the levels of plasma homovanillac acid (HVA), the major metabolite of DA, have sometimes suggested increased plasma HVA prior to treatment predicts response to neuroleptic drugs (Bowers et al. 1984; Pickar et al. 1984, 1986; Chang et al. 1988; Davila et al. 1988; Mazure et al. 1991). We found differences in plasma HVA concentrations among male and female neuroleptic-resistant or -responsive schizophrenics (defined by criteria of Kane et al. 1988; Table 2). Seven studies have found that good responders to neuroleptics have a decrease in plasma HVA concentrations during the first 4–6 weeks of treatment, but poor responders do not (see Davis et al. 1991).

Table 1. CT measures in neuroleptic-resistant and -responsive schizophrenic patients

Measure	Group: Neuroleptic-resistant ($n = 64$)	Neuroleptic-responsive ($n = 10$)
Ventricular brain ratio	6.19 ± 2.59	5.70 ± 2.86
Third ventricular width	2.31 ± .71	2.55 ± .91
Prefrontal sulcal widening	1.36 ± .53	1.53 ± .42

Table 2. Plasma HVA in neuroleptic-resistant and -responsive schizophrenic patients

		Plasma HVA (pmol/ml)		
Group	*n*	Men	*n*	Women
Neuroleptic-resistant	40	47.8 ± 22.1	23	57.1 ± 35.9
Neuroleptic-responsive	57	45.9 ± 17.6	11	54.0 ± 42.9

Dopamine Receptor Blockade and Neuroleptic Resistance

Neuroleptic resistance in schizophrenia is not due to inadequate plasma levels of antipsychotic drugs, nor is there any evidence that diminished occupancy of striatal D_2 receptors is the basis for limited response (Wolkin et al. 1989; Coppens et al. 1991; Geaney et al. 1992; Pilowsky et al. 1993). These results suggest that neuroleptic-resistant patients have a form of psychosis which is either independent of or insensitive to D_2 receptor activity and blockade.

The question of neuroleptic resistance as a discrete subtype of schizophrenia versus part of the schizophrenia or psychosis continuum has taken on new significance because of the clear evidence that up to 60% of neuroleptic-resistant schizophrenic patients will show an improvement in psychopathology if treated with clozapine, a dibenzodiazepine tricyclic drug, chemically related to loxapine and amoxapine (Kane et al. 1988; Meltzer 1991, 1992; Meltzer et al. 1989). There is also evidence that clozapine is more effective than typical neuroleptic drugs in treating positive and negative symptoms in some typical schizophrenic patients (Claghorn et al. 1987; Meltzer et al. 1989; Meltzer 1992). At the same time, clozapine is as effective as an antidepressant in patients with treatment-resistant depression (Parsa et al. 1991) and in some patients with refractory mania (Calabrese et al. 1991; Suppes et al. 1992). The effectiveness of clozapine to treat therapy-resistant psychotic depression and manic patients provides some pharmacologic evidence that there is a continuum operative across the group of psychotic disorders separated by Kraepelin into distinct subtypes. With this type of nonspecific superiority as an antipsychotic, great caution is needed in addressing the issue of whether or not the group of neuroleptic-resistant schizophrenic patients who respond to clozapine represent a unique subgroup of schizophrenia, while the group which does not respond to either neuroleptics or clozapine may be yet a third group.

The same precautions previously mentioned in the discussion of whether schizophrenic patients have had an adequate response to typical neuroleptics are relevant to the issue of an adequate response to clozapine. Unless there is adequate duration of the trial (Meltzer 1989b), adequate dose and plasma levels (Hasegawa et al. 1993), an adequate attempt to use clozapine monotherapy (Meltzer 1992), and possibly psychosocial treatments, false-negative results are likely.

The issue of whether clozapine achieves its antipsychotic effects via the same or different mechanisms than standard neuroleptic drugs is also a relevant consideration here. Until recently, it was believed that the major difference between clozapine and typical neuroleptic drugs was that the former inactivated only mesolimbic DA neurons, while the latter inactivated both mesolimbic and striatal DA neurons (Chiodo and Bunney 1983). It is now highly likely that there are much more fundamental differences between these agents, with clozapine blocking other types of DA receptors such as the D_1 and D_4 and, more importantly, a wide range of serotonin (5-HT) receptors, such as the 5-HT_{1A}, 5-HT_{2A}, 5-HT_{2C}, 5-HT_3, 5-HT_6, and 5-HT_7, some of which are not affected by the typical neuroleptic drugs (Meltzer 1991; Roth et al. 1994). These are by no means the only differences between clozapine and the typical neuroleptic drugs.

In a situation where two groups of patients have phenomenologically similar disorders, one of which responds to either of two different classes of drugs (i.e., clozapine or typical neuroleptics) and the other to only one of the drugs (i.e., clozapine), can it be concluded on this basis alone that these groups have different types of disorders, that they have fundamentally different neurochemical or structural brain abnormalities? Insofar as they both respond to the same drug, e.g., clozapine, they would appear to have similar pathophysiology. The failure of the neuroleptic-resistant group to respond to typical neuroleptic drugs suggests that some biological process which enables typical neuroleptic drugs to ameliorate positive symptoms is qualitatively absent or quantitatively deficient in these patients. This hypothesis might apply best to patients who initially respond to typical neuroleptic drugs and then lose that response. This might be due to some neuronal system which is efferent from the mesolimbic DA system being insensitive to diminished dopaminergic input while clozapine can still affect these nuclei. A second hypothesis is that neuroleptic-resistant patients have an additional nondopaminergic component which must be addressed before decreasing dopaminergic activity has a significant effect. We have suggested that this may be 5-HT_2-mediated neurotransmission (Meltzer 1989a). In this model, blocking dopaminergic activity is of some importance, but the additional effect on 5-HT_2 neurotransmission is essential in the neuroleptic-resistant patients. The latter hypothesis might apply to patients who never responded to typical neuroleptics, although some of these patients may have a DA-sensitive psychosis from the outset of their illness. These hypotheses might be pursued by studying the effect of DA agonists in treatment-resistant versus non-treatment-resistant schizophrenic patients. We hypothesize that unmedicated treatment-resistant schizophrenia cuses would have a positive response to *d*-amphetamine, while unmedicated non-treatment-resistant cases would be more likely to have an exacerbation of their psychosis.

Comparison of Neuroleptic-resistant and Nonresistant Schizophrenia

Since 1986, we have been classifying patients as neuroleptic responsive or neuroleptic resistant by the criterion of Kane et al. (1988). The majority, but not all, of the neuroleptic-resistant patients were subsequently treated with clozapine. Without considering the possibility of clozapine nonresponders being a separate subgroup, we have compared 54 neuroleptic responders and 175 neuroleptic nonresponders (criteria of Kane et al. 1988) on a variety of measures. The results will be presented in detail elsewhere (H. Y. Meltzer et al., manuscript in preparation) but a brief summary is of use here. The neuroleptic-resistant and -responsive groups differed with respect to age at index admission (33.7 ± S.D. 9.3 years and 36.2 ± S.D. 10.7 years, respectively), but there is no difference in duration of illness. Female neuroleptic-responsive patients have an older age of onset (by 5.1 years) than female neuroleptic-resistant patients. There was a trend (2.0 years) in the same direction for male patients ($p = 0.06$). Female neuroleptic-responsive patients have a significantly older age of onset than male neuroleptic-responsive patients, but there was no difference between the male and female neuroleptic-resistant patients (Table 3). Female patients were over-represented in the neuroleptic-resistant group (30%) compared with the neuroleptic-responsive group (18%)

Table 3. Age of onset in neuroleptic-resistant and -responsive schizophrenic patients

Group	*n*	Age of onset (years) Men	*n*	Women
Neuroleptic resistant	107	19.2 ± 3.7	68	20.0 ± 5.6[a]
Neuroleptic responsive	35	21.2 ± 6.8*	18	25.1 ± 8.7**

ANOVA table $F = 6.75$; $df = 3224$; $p = .002$

Source	*df*	*f*	*p*
Group	1	16.33	.001
Sex	1	7.24	.008
Group by sex	1	3.22	.07
Error	224		

*$p < .06$ versus neuroleptic-resistant men.
**$p < .004$ versus neuroleptic-resistant women and $p < .01$ versus neuroleptic-responsive men.
[a] Not significantly different from neuroleptic-resistant men.

(chi-square = 12.8, $p = 0.0001$). This may reflect the pattern of referral to this program, rather than a genuine gender difference, since there is no evidence for greater neuroleptic resistance in women than in men in the literature on schizophrenia. Rather, the reverse may be true. The two groups, because of differences in age at index admission, have a similar duration of illness, but the neuroleptic-resistant patients had more hospitalizations than the neuroleptic-responsive patients (8.8 ± 7.7 versus 6.5 ± 6.6 years, $p = 0.02$). The disorganized subtype of schizophrenia was more common in the neuroleptic-resistant group (8.3% versus 2.8%), while the residual subtype was over-represented in the neuroleptic-resistant group (11.3% versus 5.9%), but the proportions of paranoid and undifferentiated subtypes were not significantly different.

We found evidence of impaired premorbid function with regard to peer group friendships and activities and interests in both groups, the neuroleptic-responsive group being somewhat more impaired in early years (up to age 15) while the neuroleptic-resistant group showed more dysfunction during the period prior to first psychiatric contact. Whelton et al. (1992) found that psychosocial dysfunction, including delayed sexual behavior, during developmental years, predicted persistent negative symptoms and global assessment following a 6 month neuroleptic trial. There was no difference in the frequency of family histories of psychotic illness. The neuroleptic-resistant patients married less frequently than the neuroleptic-responsive patients, but the two groups did not differ with regard to highest education attained or occupational status.

The neuroleptic-resistant patients tended to have higher levels of negative symptoms and poorer quality-of-life ratings as measured by the SANS (Andreasen 1983) and Heinrichs et al. scale (1984). The two groups did not differ on BPRS total, positive, or negative symptoms or on SADS-C Disorganization at admission (data not presented).

The results reported here are in general agreement with those in the literature with some exceptions. Bartkó et al. (1990) also found younger age of onset in neuroleptic-resistant patients, while Kolakowska et al. (1985) and Nimgaonkar et al. (1988) did not. The number of subjects was quite small in these two studies. Worse premorbid function has previously been reported to be present in neuroleptic-resistant patients (Kolakowska et al. 1985; Bartkó et al. 1990; Keefe et al. 1990) as we found here, but only in the prodromal period. This issue requires further study. We found no evidence that neuroleptic-resistant patients had a positive family history of schizophrenia more frequently than responders, and neither did Nimgaonkar et al. (1988), whereas Silverman et al. (1987) did find an association between the two. Silverman et al. (1987) used the family study method, which may be more reliable than the family history method. The finding of higher levels of negative symptoms in neuroleptic-resistant patients is consistent with that of Crow (1980), who suggested greater negative symptoms were characteristic of neuroleptic-resistant patients. Kolakowska et al. (1985) reported that neuroleptic-resistant patients were more likely to have greater amounts of both positive and negative symptoms.

The most intriguing data are the similar age of onset in neuroleptic-resistant female patients and neuroleptic-responsive female and male patients compared with the expected large differences in age of onset between males and females. The 5-year difference in age of onset between neuroleptic-resistant and neuroleptic-responsive female patients is rather remarkable. The number of women in this study is rather small, however, and so these results must be confirmed in additional studies. The few other differences between neuroleptic-responsive and -resistant schizophrenic patients directly reflect the lack of response to treatment (e.g., more hospitalizations, more negative symptoms, poor quality of life). There was no evidence from the CT studies that cerebral atrophy or cognitive impairment as measured by a battery of cognitive tests (Kenny and Meltzer 1991) differed between the groups. It would be premature to conclude on the basis of the age-of-onset data alone that neuroleptic-resistance represents a separate subtype of schizophrenia. The combination of a process which begins 5 years earlier on average and a poorer response to neuroleptics suggests that the neuroleptic-resistant female patients, at least, might represent a distinct subtype. We are currently investigating whether the neuroleptic-resistant patients showed an inadequate response to neuroleptics at the time of onset or developed it subsequently. Earlier age of onset might lead to poorer outcome because of a quantitatively more malignant process but not necessarily a qualitatively different process. Earlier age of onset could also cause poorer outcome simply because fewer years of normal development before psychosis leads to greater interference in psychosexual maturation and social function. However, there was no significant difference in age of onset between neuroleptic-responsive male patients (21.2 ± 0.90 years) and neuroleptic-resistant female patients (20.0 ± 0.65; $p = 0.29$), suggesting that age of onset per se does not lead to neuroleptic resistance.

Conclusions

Pharmacologic challenge and treatment studies may be of some value in answering complex nosologic issues such as whether schizophrenia and mood disorders are part of a continuum or whether schizophrenia consists of more than one subtype. Pharmacologic challenge studies of the dopaminergic system have not provided evidence that permits either rejection of the continuum hypothesis or identification of distinct subtypes of schizophrenia. However, studies of serotonergic dysfunction are more suggestive of distinctive subtypes of schizophrenia and tend to be inconsistent with a continuum hypothesis. Treatment-response studies tend to be partial supportive of a continuum process, to the extent that both typical and atypical antipsychotics have a clear antipsychotic effect in both groups of disorders. Evidence in support of the hypothesis that treatment-resistant schizophrenia, defined by failure to respond to adequate trials of typical neuroleptic drugs, is a separate subtype is accumulating. Age-of-onset data suggest that this may be especially true of female patients.

References

Andreasen NC (1983) The scale for the assessment of negative symptoms (SANS). University of Iowa, Iowa City
Angrist B, Peselow E, Rubinstein M, Corwin J, Rotrosen J (1982) Partial improvement in negative schizophrenic symptoms after amphetamine. Psychopharmacology (Berlin) 78:128–130
Asnis GM, Ryan ND (1983) The psychoneuroendocrinology of schizophrenia. In: Rifkin A (ed) Schizophrenia and affective disorders: biology and drug treatment. John Wright, Boston
Bartkó G, Frecska E, Zádor G, Herczy I (1989) Neurological features, cognitive impairment and neuroleptic response in schizophrenic patients. Schizophr Res 2: 311–313
Bartkó G, Frecska E, Horvath S, Zador G, Arato M (1990) Predicting neuroleptic response from a combination of multilevel variables in acute schizophrenic patients. Acta Psychiatr Scan 82:408–412
Bleich A, Brown S-L, Kahn R, Van Praag HM (1988) The role of serotonin in schizophrenia. Schizophr Bull 14:297–316
Bowers MB Jr, Swigar ME, Jatlow PI, Golcocchea N (1984) Plasma catcholamine metabolites and early response to haloperidol. J Clin Psychiatry 45:248–251
Bowers MB, Mazure CM, Nelson CJ, Jatlow PI (1990) Psychotogenic drug use and neuroleptic response. Schizophr Bull 16:81–85
Brenner HD, Dencker SJ, Goldstein MJ, Hubbard JW, Keegan DL, Kruger G, Kulhanek F, Liberman RP, Malm V, Midha KK (1990) Defining treatment refractoriness in schizophrenia. Schizophr Bull 17:551–562
Brown GM, Cleghorn JM, Kaplan RD, Szechtman H, Mitton J (1988) Longitudinal growth hormone studies in schizophrenia. Psychiatr Res 24:123–136
Brown WA, Herz LR (1988) Neuroleptic response as a nosologic device. In: Tsuang MT, Simpson JG (eds) Handbook of schizophrenia: noslogy, epidemiology and genetics, vol 3. Elsevier Science, New York, pp 139–149
Buckley P, Thompson P, Way L, Meltzer HY (1994) Substance abuse among patients with treatment-resistant schizophrenia: characteristics and implications for clozapine therapy. Am J Psychiatry (in press)
Caccavelli L, Cussac D, Pellegrini I, Andenot V, Jaquet P, Enjalbert A (1992) D_2 dopaminergic receptors: normal and abnormal transduction mechanism. Hormone Res 38:78–83.
Calabrese JR, Meltzer HY, Markovitz PJ (1991) Clozapine prophylaxis in rapid cycling: bipolar disorder. J Clin Psychopharmacol 11:396–397
Carpenter WT, McGlashan TH, Strauss JS (1977) The treatment of acute schizophrenia without drugs: an investigation of some current assumptions. Am J Psychiatry 134: 14–20
Cesarec A, Nyman AK (1985) Differential response to amphetamine in schizophrenia. Acta Psychiatr Scand 71:523–528
Chang W-H, Chen T-Y, Lee C-F, Hung J-C, Hu W-H, Yeh E-K (1988) Plasma homovanillac acid levels and subtyping of schizophrenia. Psychiatr Res 23:239–244
Chiodo LA, Bunney BS (1983) Typical and atypical neuroleptics: differential effects of chronic administration on the activity of A9 and A10 midbrain dopaminergic neurons. J Neurosci 3:1607–1619

Claghorn J, Honigfeld G, Abuzzahab FS, Wang R, Steinbook R, Tuason V, Klerman G (1987) The risks and benefits of clozapine versus chlorpromazine. J Clin Psychopharmacol 7:377–384
Cleghorn JM, Szechman H, Garnett ES, Nehmias C, Brown GM, Kaplan RD, Franco S (1991) Apomorphine effects on brain metabolism in neuroleptic-naive first-episode schizophrenic patients and normal controls. Psychiatr Res 40:135–153
Coppens HJ, Slooff CJ, Poans AMJ, Weigman T, Vaalburg W, Korf J (1991) High central D_2 dopamine receptor occupancy as assessed with positron emission tomography in medicated but therapy-resistant schizophrenic patients. Biol Psychiatry 29:629–634
Crayton JW, Stålberg E, Hilton-Brown P (1977) The motor unit in psychotic patients: a single-fibre EMG study. J Neurol Neurosurg Psychiatry 40:455–463
Crow TJ (1980) Molecular pathology of schizophrenia: more than one disease process? Br Med J 280-66-68
Crow TJ (1990) The continuum of psychosis and its genetic origins. The sixty-fifth Maudsley lecture. Br J Psychiatry 157:781–782
Crow TJ, Ferrier IN, Johnstone EC (1986) The two-syndrome concept and neuroendocrinology of schizophrenia. Psychiatr Clin North Am 9:99–113
Csernansky JE, Kaplan J, Hollister LE (1985) Problems in the classification of schizophrenics as neuroleptic responders and nonresponders. J Nerv Mental Dis 173: 325–331
Daniel DG, Berman KP, Weinberger DR (1989) The effect of apomorphine on regional cerebral blood flow in schizophrenia. J Neuropsychiatr Clin Neurosci 1:377–384
Daniel DG, Weinberger DR, Jones DW, Zigun JR, Coppola R, Handel S, Bigelow LB, Goldberg TE, Berman KF, Kleinman JE (1991) The effect of amphetamine on regional cerebral blood flow during cognitive activation in schizophrenia. J Neurosci 11:1907–1917
Davila R, Manero E, Zumarraga M, Andia I, Schweitzer JW, Friedhoff AJ (1988) Plasma homovanillac acid as a predictor of response to neuroleptics. Arch Gen Psychiatry 45:564–567
Davis KL, Kahn RS, KO G, Davidson M (1991) Dopamine and schizophrenia: a reivew and reformulation. Am J Psychiatry 148:1474–1486
Delgado PL, Charney DS, Price LH, Landis H, Heninger GR (1990) Neuroendocrine and behavioral effects of dietary tryptophan restriction in healthy subjects. Life Sci 45:2323–2332
Ferrier IN, Johnstone EC, Crow TJ (1984) Hormonal effects of apomorphine in schizophrenia. Br J Psychiatry 144:349–357
Friedman L, Lys C, Schulz SC (1992) The relationship of structural brain imaging parameters to antipsychotic treatment response: a review. J Psychiatr Neurosci 17:42–54
Garver DL, Kelley K, Fried KA, Magnusson M, Hirschowitz J (1988) Drug response patterns as a basis of nosology for the mood-incongruent psychoses (the schizophrenias). Psychol Med 18:873–885
Geaney DP, Ellis PM, Soper N, Shepstone BJ, Cowen PJ (1992) Single photon emission tomography assessment of cerebral dopamine D_2 receptor blockade in schizophrenia. Biol Psychiatry 32:293–295
Goldberg TE, Bigelow LB, Weinberger DR, Daniel DG, Kleinman JE (1991) Cognitive and behavioral effects of the coadministration of dextroamphetamine and haloperidol in schizophrenia. Am J Psychiatry 148:78–84
Goldstein BJ (1970) Comparative efficacy of thiothixene capsules and concentrate in acutely schizophrenic patients. Psychosomatics 11:434–437
Gruen PH, Sashar EJ, Langer G et al (1978) Prolactin responses to neuroleptics in normal and schizophrenic subjects. Arch Gen Psychiatry 35:108–118

Hasegawa M, Gutierrez-Esteinou R, Way L, Meltzer HY (1993) Relationship between clinical efficacy and clozapine plasma concentrations in schizophrenia: effect of smoking. J Clin Psychopharmacol 13:383–390
Heinrichs DW, Hanlon TE, Carpenter WT Jr (1984) The quality-of-life scale: an instrument for rating the schizophrenic deficit syndrome. Schizophr Bull 10:388–306
Hirschowitz J, Zemlan FP, Garver DL (1982) Growth hormone levels and lithium ratios as predictors of success of lithium therapy in schizophrenia. Am J Psychiatry 139: 646–649
Hirschowitz J, Zemlan FP, Hitzeman RJ, Fleischmann RL, Garver DL (1986) Growth hormone response to apomorphine and diagnosis: a comparison of three diagnostic systems. Biol Psychiatry 21:445–454
Hollister EE, Overall JE, Bennett JL, Kimball I Jr, Shelton J (1967) Specific therapeutic actions of acetophenazine, perphenazine and benzquinamole in newly admitted schizophrenic patients. Clin Pharmacol Ther 8:249–255
Iqbal N, Goldsant LA, Wetzler S, Schwarz BJ, Van Praag HM (1991) Serotonin and schizophrenia. Psychiatr Ann 23:186–193
Jody D, Lieberman JA, Geisler S, Szymanski S, Alvir JM (1990) Behavioral response to methylphenidate and treatment outcome in first-episode schizophrenia. Psychopharmacol Bull 26:224–230
Kahn RS, Siever L, Gabriel S, Amin F, Stern RG, Dummont K, Apter S, Davidson M (1992) Serotonin function in schizophrenia: effects of m-chlorophenylpiperazine in schizophrenic patients and healthy subjects. Psychiatr Res 43:1–12
Kane J, Honigfeld G, Singer J, Meltzer HY, the Clozaril Collaborative Study Group (1988) Clozapine for the treatment-resistant schizophrenic: a double-blind comparison with chlorpromazine. Arch Gen Psychiatry 45:789–796
Kay SR, Fiszbein A, Opler LA (1987) The positive and negative symptoms scale (PANSS) for schizophrenia. Schizophr Bull 12:578–602
Kay SR, Opler LA, Lindenmayer J-P (1988) Reliability and validity of the positive and negative syndrome scale for schizophrenics. Psychiatr Res 23:99–100
Keefe RSE, Mohn RC, Silverman JM et al (1990) Characteristics of Kraepelinian schizophrenia an their relation to premorbid sociosexual functioning. In: Agrist B, Schulz SC (eds) The neuroleptic nonresponsive patient: characterization and treatment. American Psychiatric Press, Washington DC
Keks NA, Copolov DL, Kulkarni Jm et al (1990) Basal and haloperidol-stimulated prolactin in neuroleptic-free men with schizophrenia defined by eleven diagnostic systems. Biol Psychiatry 27:1203–1215
Keks NA, McKenzie DP, Low LH, McGorry PD, Hill C, Kulkarni J, Bingh BS, Copolov DL (1992) Multidiagnostic evaluation of prolactin response to haloperidol challenge in schizohrenia: maximal blunting in Kraepelinian patients. Biol Psychiatry 32:426–437
Kendell RE (1975) The role of diagnosis in psychiatry. Blackwell Scientific, Oxford
Kenny J, Meltzer HY (1991) Attention of higher cortical functions in schizohprenia. J Neuropsychiatr Clin Neurosci 3:269–275
Kolakowska T, Williams AO, Andern M, Reveley MA, Jambor K, Gelder MG, Mandelbrote BM (1985) Schizophrenia with good and poor outcome. I. Early clinical features, response to neuroleptics and signs of organic dysfunction. Br J Psychiatry 146:229–246
Lee HS, Bastani B, Friedman L, Ramirez L, Meltzer HY (1991) Effect of the serotonin agonist, MK-212, on body temperature in schizophrenia. Biol Psychiatr 31:460–470
Lerer B, Ran A, Blacker M, Silver H, Weller MPI, Drummer D, Ebstein B, Calev A (1988) Neuroendocrine responses in chronic schizophrenia: evidence for serotonergic dysfunction. Schizophr Res 1:405–410
Lesch K-P, Mayer S, Disselkamp-Tietze J, Hoh A, Weissman M, Osterheider M, Schulte HM (1990) 5-HT_{1A} receptor responsivity in unipolar depression: evaluation

of ipsapirone-induced ACTH and cortisol secretion in patients and controls. Biol Psychiatry 28:620–268

Liddle PF (1995) Associations between regional brain abnormalities and specific syndromes in schizophrenia. In: Häfner H, Gattaz WF (eds) Search for the causes of schizophrenia. III. Springer, Bevlin Heidelberg New York (in press)

Lieberman JA, Kane JM, Gadaleta D, Brenner R, Lesser MS, Kinon B (1984) Methylphenidate challenge as a predictor of relapse in schizophrenia. Am J Psychiatry 141:633–638

Lieberman JA, Kane JM, Sarantakos S, Gadaleta D, Woerner M, Alvir J, Ramos-Lorenzi J (1987) Prediction of relapse in schizophrenia. Arch Gen Psychiatry 44: 597–603

Losonczy MF, Song IS, Mohs RC, Mathe AA, Davidson M, Davis BM, Davis KL (1986) Correlates of lateral ventricular size in chronic schizohprenia. II. Biological correlates. Am J Psychiatry 143:1113–1118

Malas KL, van Kammen DP, de-Fraites EA, Brown GM, Gold PE (1987) Reduced growth hormone response to apomorphine in schizophrenic patients with poor premorbid social functioning. J Neural Transm 69:319–324

Mazure CM, Nelson JC, Jatlow PI, Bowers MB (1991) Plasma free homovanillic acid (HVA) as a predictor of clinical response in acute psychosis. Biol Psychiatry 30: 475–482

Meltzer HY (1973) Skeletal muscle abnormalities in patients with affective disorders. J Psychiatr Res 10:43–57

Meltzer HY (1975) Neuromuscular abnormalities in the major mental illnesses. I. Serum enzyme studies. In: Freedman DX (ed) Biology of the major psychoses, research publications association of the research in nervous mental disorders, vol. 54. Raven, New York

Meltzer HY (1976) Neuromuscular dysfunction in schizophrenia. Schizophr Bull 2: 106–135

Meltzer HY (1979) Clinical evidence for multiple dopamine receptors in man. Commun Psychopharmacol 3:457–470

Meltzer HY (1982) Dopamine autoreceptor stimulation: clinical significance. Pharmacol Biochem Behav 17:1–10

Meltzer HY (1985) Dopamine and negative symptoms in schizophrenia: critique of the type I–type II hypothesis. In: Alpert M (ed) Controversies in schizophrenia: changes and constancies. Guilford, London, pp 110–136

Meltzer HY (1987) Biological studies in schizophrenia. Schizophr Bull 13:77–111

Meltzer HY (1989a) Clinical studies on the mechanism of action of clozapine: the dopamine-serotonin hypothesis of schizophrenia. Psychopharmacology 99:S18–S27

Meltzer HY (1989b) Letter to the Editor. Duration of a clozapine trial in neuroleptic-resistant schizophrenia. Arch Gen Psychiatry 46:672

Meltzer HY (1991) Atypical antipsychotic drugs: the 5-HT_2/ DA_2 ratio: Chap. 24. In: 5-Hydroxytryptamine in psychiatry. In: Sandler M, Coppen A, Harnett S (eds) A spectrum of ideas. Oxford University Press, London

Meltzer HY (1992) Dimensions of outcome with clozapine. Br J Psychiatry 160 [Suppl]: 54–59

Meltzer HY, Crayton JW (1975) Neuromuscular abnormalities in the major mental illnesses. II. Muscle fiber and subterminal motor nerve abnormalities. In: Freedman DX (ed) Biology of the major psychoses, research publications association of the research in nervous mental disorders, vol. 54. Raven, New York

Meltzer HY, Fang VS (1976) The effect of neuroleptics on serum prolactin in schizophrenic patients. Arch Gen Psychiatry 33:279–286

Meltzer HY, Lowy MT (1987) The serotonin hypothesis of depression. In: Meltzer HY (ed) Psychopharmacology, the third generation of progress. Raven, New York

Meltzer HY, Stahl SM (1976) The dopamine hypothesis of schizophrenia: a review. Schizophr Bull 2:19–76

Meltzer HY, Busch D, Fang VS (1981) Hormone, dopamine receptors and schizophrenia. Psychoneuroendocrinology 6:17–36
Meltzer HY, Perline R, Lewine R (1982) Biological studies of DSM-III psychotic disorders. 1. Platelet measures and apomorphine-induced growth hormone response. J Nerv Mental Dis 170:758–765
Meltzer HY, Kolakowska T, Fang VS, Fogg L, Robertson A, Lewine R, Strahilevitz M, Busch D (1984a) Growth hormone and prolactin response to apomorphine in schizophrenia and the affective disorders: relation to duration of illness and depressive symptoms. Arch Gen Psychiatry 41:572–579
Meltzer HY, Uberkoman-Wiita B, Robertson A, Tricou BJ, Lowy M, Perline R (1984b) Effect of 5-hydroxytryptophan on serum cortisol levels in the major affective disorders and normal controls. I. Enhanced response in depression and mania. Arch Gen Psychiatry 41:366–374
Meltzer HY, Sommers AA, Luchins DJ (1986) The effect of neuroleptics and other psychotropic drugs on negative symptoms in schizophrenia. J Clin Psychopharmacol 6:329–338
Meltzer HY, Bastani B, Kwon KY, Ramirez LF, Burnett S, Sharpe J (1989) A prospective study of clozapine in treatment-resistant patients. I. Preliminary report. Psychopharmacology 99 [Suppl]:S68–S72
Meltzer HY, Maes M, Lee MA (1993) The cimetidine-induced increase in prolactin secretion in schizophrenia: effect of clozapine. Psychopharmacology 112:S95–S104
Muller-Spahn F, Ackenheil M, Albus M, May G, Naber D, Welter D, Zander K (1984) Neuroendocrine effects of apomorphine in chronic schizophrenic patients under long-term neuroleptic therapy and after drug withdrawal: relations to psychopathology and tardive dyskinesia. Psychopharmacology (Berlin) 84: 436–440
Nimgaonkar VL, Wesseley S, Tune LE, Murray RM (1988) Response to durgs in schizophrenia: the influence of family history, obstetric complications and ventricular enlargement. Psychol Med 18:583–592
Overall JE, Gorham D (1962) The brief psychiatry rating scale. Psychol Rep 10: 149–165
Overall JE, Rhodes HM (1982) Refinement of phenomenological classification in clinical psychopharmacology research. Psychopharmacology 77:24–30
Pandurangi AK, Goldberg SC, Brink DD, Hill MH, Gulati AN, Hamer RM (1989) Amphetamine challenge test, response to treatment, and lateral ventricle size in schizophrenia. Biol Psychiatry 25:207–214
Parsa MA, Ramirez LF, Loula EC, Meltzer HY (1991) Effect of clozapine on psychotic depression and parkinsonism. J Clin Psychopharmacol 11:330
Pickar D, Labraca A, Linnoila M, Roy A, Hommer D, Everett D, Paul SM (1984) Neuroleptic-induced decrease in plasma homovanillac acid and antipsychotic activity in schizophrenic patients. Science 225:954–957
Picker D, Labarca R, Doran AR, Wolkowitz OM, Roy A, Brier A, Linnoila M, Paul SM (1986) Longitudinal measurement of plasma homovanillac acid levels in schizophrenic patients. Arch Gen Psychiatry 43:669–676
Pilowsky LS, Costa DC, Ell PJ, Murray RM, Verhoeff NPLG, Kerwin RW (1993) Antipsychotic medication, D_2 dopamine receptor blockade and clinical response: a ^{123}I IBZMSPECT (single photon emission tomography) study. Psychol Med 23: 791–797
Raz S, Raz N (1990) Structural brain abnormalities in the major psychoses: a quantitation review of the evidence from computerized imaging. Psychol Bull 108:93–198
Roth BL, Craigo SC, Choudhary MS, Uler A, Monsma FJ, Shen Y, Meltzer HY, Sibley DR (1994) Binding of typical and atypical antipsychotic agents to 5-hydroxytryptamine$_6$ (5-HT$_6$) and 5-hydroxtryptamine$_7$ (5-HT$_7$) receptors. J Pharmacol Exp Ther (in press)

Sajatovic M, Meltzer HY (1993) The effect of short-term electroconvulsive treatment plus neuroleptics in treatment-resistant schizophrenia and schizoaffective disorder. Convulsive Ther 9:167–175
Sautter F, Garver DL, Zemlan FP, Hirschowitz J (1987) Growth hormone response to apomorphine and family patterns of illness. Biol Psychiatr 22:717–724
Sautter F, McDermott B, Garver D (1993) Familial differences between rapid neuroleptic response psychosis and delayed neuroleptic response psychosis. Biol Psychiatry 33:15–21
Schulz SC, Conley RR, Kahn EM, Alexander J (1988) Nonresponders to neuroleptics: a distinct subtype. In: Schulz SC, Tamminga CA (eds) Schizophrenia: scientific progress. Oxford University Press, New York
Seibyl JP, Satel SL, Anthony D, Southwick SM, Krystal JH, Charney DS (1993) Effects of cocaine on hospital course in schizophrenia. J Nerv Ment Dis 181:31–37
Sharma RP, Javaid JI, Pandey GN, Easton M, Davis JM (1990) Pharmacological effects of methylphenidate on plasma homovanillic acid and growth hormone. Psychiatr Res 32:9–17
Sharma RP, Javaid JI, Pandey GN, Janicak PG, David JM (1991) Behavioral and biochemical effects of methylphenidate in schizophrenic and non-schizophrenic patients. Biol Psychiatry 30:459–466
Silverman JM, Mohs RC, Davidson M, Losonczy MF, Keefe RS, Breitner JC (1987) Familial schizophrenia and treatment response. Am J Psychiatry 144:1271–1276
Strauss JS, Carpenter WT Jr (1972) The prediction of outcome in schizophrenia. I. Characteristics of outcome. Arch Gen Psychiatry 27:739–746
Strauss JS, Carpenter WT Jr (1978) The prognosis of schizophrenia: rationale for a multidimensional concept. Schizophr Bull 4:56–67
Suppes T, McElroy SL, Gilbert J, Dessain EC, Cole JO (1992) Clozapine in treatment of dysphoric mania. Biol Psychiatry 32:270–280
Thompson PA, Buckley PF, Meltzer HY (1994) The brief psychiatric rating scale: effect of scaling system on clinical response assessment. J Clin Psychopharmacol (in press)
Van Kammen DP, Boronow JJ (1988) Dextroamphetamine diminishes negative symptoms in schizophrenia. Int Clin Psychopharmacol 3:111–121
Van Kammen DP, Docherty JP, Marder SR, Bunney WE Jr (1981) Acute amphetamine response predicts antidepressant and antipsychotic responses to lithium carbonate in schizophrenic patients. Psychiatr Res 4:313–325
Van Kammen DP, Docherty JP, Bunney WE Jr (1982a) Prediction of early relapse after pimozide discontinuation by response to d-amphetamine during pimozide treatment. Biol Psychiatry 17:233–242
Van Kammen DP, Sternberg DE, Hare TA, Waters RN, Bunney WE Jr (1982b) CSF levels of γ-aminobutyric acid in schizophrenia: low values in recently ill patients. Arch Gen Psychiatry 39:91–97
Whalley LJ, Christie JE, Brown S, Arbuthnott GW (1984) Schneider's first-rank symptoms of schizophrenia. An association with increased growth hormone response to apomorphine. Arch Gen Psychiatry 41:1040–1043
Whelton CL, Cleghorn JM, Atley S, Durocher GJ, MacCrimmon D (1992) Developmental and neurologic correlates of treatment response in schizophrenia. J Psychiatr Neurosci 17:15–22
Wolf ME, Bowie L, Keener S, Mosniam AD (1982) Prolactin response in tardive dyskinesia. Biol Psychiatry 17:485–490
Wolkin A, Angrist B, Wolf A, Brodie J, Wolkin B, Jaeger J, Cancro R, Rotrosen J (1987) Effects of amphetamine on local cerebral metabolism in normal and schizophrenic subjects as determined by positron emission tomography. Psychopharmacology (Berlin) 92:241–246
Wolkin A, Barouche F, Wolf AP, Rotrosen J, Fowler JS, Shine C-Y, Cooper TB, Brodie JD (1989) Dopamine blockade and clinical response: evidence for two biological subgroups of schizophrenia. Am J Psychiatry 146:905–908

Young MA, Meltzer HY (1980) The relationship of demographic, clinical and outcome variables to neuroleptic treatment requirements. Schizophr Bull 6:88–101
Zemlan FP, Hirschowitz J, Garver DL (1986a) Relation of clinical symptoms to apomorphine-stimulated growth hormone release in mood-incongruent psychotic patients. Arch Gen Psychiatry 43:1162–1167
Zemlan FP, Hirschowitz J, Sautter F, Garver DL (1986b) Relationship of psychotic symptom clusters in schizophrenia to neuroleptic treatment and growth hormone response to apomorphine. Psychiatr Res 18:239–255

Psychotic Continuum or Distinct Entities: Perspectives from Family Studies*

M. T. Tsuang[1], J. C. Simpson[2], and J. A. Fleming[3]

There is a continuing controversy over the continuum versus categorical models for major psychoses. As stated by Crow (1990a), the continuum theory holds that "the recurrent psychoses, rather than, as Kraepelin supposed, constituting 2 major entities, manic depressive illness and schizophrenia, as separate diseases, may be distributed along a continuum that extends from unipolar depressive illness through bipolar and schizoaffective psychosis to schizophrenia with increasing severities of defect state" (p. 401). Furthermore, the continuum is hypothesized to have a genetic basis (Crow 1990a,b, 1991), which should therefore be reflected in family studies of psychotic patients.

In this paper we will present and discuss results of family studies of psychotic disorders, with special emphasis on schizoaffective disorder and its relationship to schizophrenia and affective disorder. The special importance of schizoaffective disorder in this regard is that it has long presented a puzzle to traditional Kraepelinian concepts and does not fit neatly into the two-entities model of discrete schizophrenic and affective disorders. In particular, the existence of substantial numbers of patients with mixed schizophrenic and affective features provides an empirical base for development of theoretical approaches that emphasize the continuity between typical schizophrenia and typical affective disorder. It is therefore crucial to examine carefully and critically the empirical results in this area to determine to what extent they support either the continuum or the Kraepelinian approach.

We have previously reported on an extensive family study of schizophrenia and affective disorders among first-degree relatives of probands with schizo-

---

*Preparation of this article was supported by the National Institute of Mental Health Grants 1 R01MH41879-01, 5 U01MH46318-02 and 1 R37MH43518-01 to Dr. Tsuang, and by the Medical Research and Health Services Research and Development Services of the Department of Veterans Affairs

[1] Stanley Cobb Professor of Psychiatry and Director, Harvard Institute of Psychiatric Epidemiology and Genetics, Harvard Departments of Psychiatry and Epidemiology and Brockton/West Roxbury VA Medical Center; Head, Department of Psychiatry, Harvard Medical School at Massachusetts Mental Health Center, 74 Fenwood Road, Boston, MA 02115, USA
[2] Program Coordinator, Brockton/West Roxbury VA Medical Center; Instructor, Department of Psychiatry, Harvard Medical School
[3] Health Statistician, Brockton/West Roxbury VA Medical Center; Principal Associate, Department of Psychiatry, Harvard Medical School

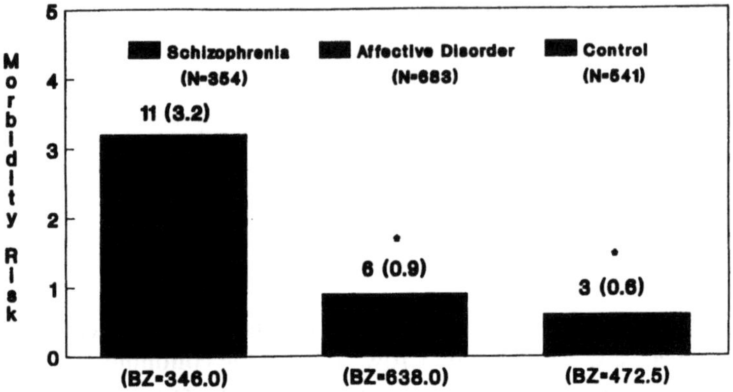

Fig. 1. Morbidity risk of schizophrenia for relative's of patients with schizophrenia, affective disorder, and control groups (*BZ*, Berugsziffer, age-adjusted size of the sample based on a risk period of 15–39 years of age for schizophrenia). $^{*}p < .05$ (comparison of schizophrenia with affective disorder and control groups)

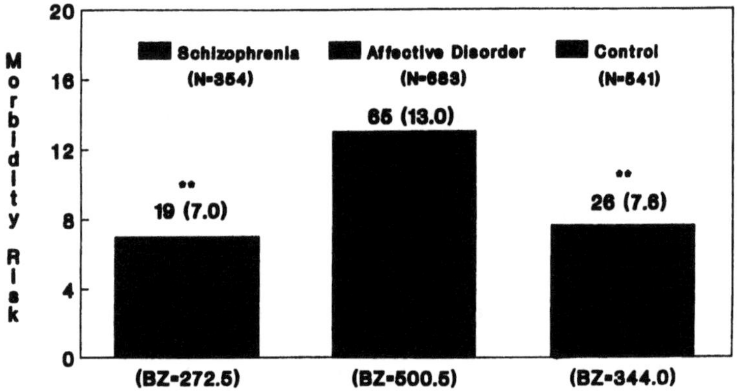

Fig. 2. Morbidity risk of affective disorder for relatives of patients with schizophrenia, affective disorder, and control groups. $^{**}p < .01$ (comparison of affective disorder with schizophrenia and control groups)

phrenia and affective disorders (Tsuang et al. 1980). Our data, based on blind diagnostic assessment of the relatives, support the distinction between schizophrenia and affective disorders from a familial perspective. As shown in Fig. 1, the age-adjusted morbidity risk of narrowly defined schizophrenia in the relatives of schizophrenics (3.2%) was significantly higher ($p < .05$) than the morbidity risk of 0.9% in the relatives of patients with major affective disorder. On the other hand, the comparison with relatives of surgical controls did not support an increased risk of schizophrenia in relatives of affective disorder patients (see Fig. 1).

As shown in Fig. 2, the morbidity risk of major affective disorder in the relatives of affective disorder patients (13.0%) was significantly higher ($p <$

.01) than in the relatives of schizophrenics (7.0%). Furthermore, there was no evidence of an increased risk for affective disorder in relatives of schizophrenics when compared with relatives of surgical controls (see Fig. 2).

Although these findings support the traditional (Kraepelinian) model of psychotic illness, they do not address directly the critical issue of the boundary between schizophrenia and affective disorder. As a number of investigators have pointed out (e.g., Kendell and Gourlay 1970; Angst et al. 1979a,b; Gershon et al. 1988; Crow 1990a), schizoaffective and other "cases in between" schizophrenia and affective disorder are of crucial importance in explaining the distribution and etiology of the major psychoses. Angst et al. (1979b) stated this viewpoint succinctly: "The sphinx of schizoaffective psychoses may be the key for an integrative genetic theory of schizophrenia and affective disorders" (p. 164).

As an early example of the use of schizoaffective disorder to address the relationship between schizophrenia and affective disorder from a familial perspective, Tsuang (1979) compared observed and expected numbers of sib pairs, where each sib had a diagnosis of schizophrenia, schizoaffective disorder, or affective disorder, and where expected numbers were computed based on a hypothesis of genetic independence between these disorders. The results showed a significant deficiency only in the observed number of same-diagnosis schizoaffective pairs, suggesting that schizophrenia and affective disorder are genetically distinct, whereas schizoaffective disorder is not. To test whether schizoaffective disorder is a variant of affective disorder or schizophrenia, further comparisons of observed and expected numbers were made using pairs with one sib schizoaffective and the other sib having either schizophrenia or affective disorder. The results of these comparisons suggested that schizoaffective disorder is genetically heterogeneous, with both affective and schizophrenic subtypes, as well as an undifferentiated subtype for patients with pronounced mixed features or insufficient information to permit subclassification.

As reviewed by Gershon et al. (1988), a number of investigators have examined morbidity risks in first-degree relatives of schizoaffective patients as a means of understanding this boundary area of psychosis. The results from our own investigation have been fairly typical of results obtained by other research groups and are represented here by Figs. 3 and 4. For purposes of this discussion of the Iowa Non500 data, I will use the term "schizoaffective" in the broad sense to refer to a heterogeneous group of patients with clinical diagnoses of schizophrenia who did not meet the Washington Univeristy criteria (Feighner et al. 1972) for schizophrenia, mania, or depression. With respect to the morbidity risk for schizophrenia, schizoaffective disorder is clearly intermediate between schizophrenia and affective disorders (Fig. 3). Similar results have been reported by Mendlewicz et al. (1980), Baron et al. (1982), and Guze et al. (1983).

In contrast, the morbidity risk for affective disorders in relatives of schizoaffective probands in the Iowa 500/Non500 data was almost identical to that in relatives of schizophrenics and significantly smaller than the morbidity risk in relatives of manic and depressed probands (see Fig. 4). A number of other

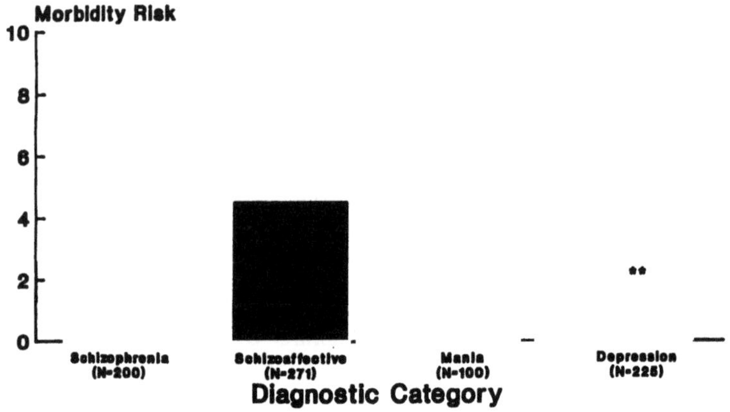

Fig. 3. Morbidity risk of schizophrenia for relatives of patients with schizoaffective disorder compared with major psychoses. $^{**}p < .01$ (comparison with schizoaffective disorder)

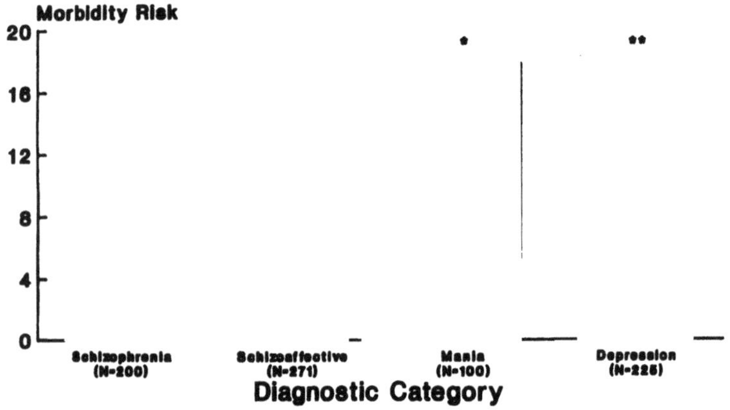

Fig. 4. Morbidity risk of affective disorder for relatives of patients with schizoaffective disorder compared with major psychoses. $^{*}p < .05$; $^{**}p < .01$ (comparison with schizoaffective disorder)

investigators (see Mendlewicz et al. 1980; Baron et al. 1982; Gershon et al. 1988; and Guze et al. 1983) have found evidence that the morbidity risk of affective disorder in relatives of schizoaffective probands is increased in comparison to the risk in relatives of schizophrenic probands, although the results have not always been consistent. For example, Gershon et al. (1988) reported that the morbidity risk of bipolar disorder was greater in relatives of chronic schizoaffective patients than in relatives of schizophrenics, but also that the reverse pattern was found with the morbidity risk of unipolar disorder in relatives.

Some differences in results from family studies are to be expected, given the differences between studies in how these disorders are defined in their study samples. In particular, results can be expected to be affected by how broadly or narrowly one defines schizophrenia and affective disorder (and hence, as a residual category, schizoaffective disorder), as well as by the selection criteria for the study population and probably various other demographic and other risk factors.

To examine these issues in more detail, some investigators have undertaken analyses of morbidity risks in subtypes of schizoaffective disorder. For example, Angst et al. (1979b) subdivided their schizoaffective proband sample using the following characteristics: sex, age at first episode, number of episodes, and various psychopathological subtypes including manic vs. nonmanic, and schizo-dominant vs. affect-dominant based on longitudinal course. There was no consistent evidence that distinct homogeneous subgroups could be identified based on their family data, although some results were in the expected direction. For example, the morbid risk of schizophrenia was 5.9% in relatives of schizo-dominant (combined with undifferentiated) schizoaffective patients and 4.6% in relatives of affect-dominant patients (n.s. difference), whereas the morbid risk of affective disorder was non-significantly greater in the affect-dominant subgroup (7.8%, vs. 5.8% for relatives of schizo-dominant probands). The authors did not report separate results for schizo-dominant probands excluding those who were classified as undifferentiated, and it is possible that the inclusion of the undifferentiated group somewhat diminished differences between the schizo- and affect-dominant subgroups. For example, in a later study that has figured prominently in the literature concerning the continuum hypothesis (Angst and Scharfetter 1990, cited by Crow 1990a,b), the ratio of morbid risks for schizophrenia to affective psychosis increased from 0.92 in affect-dominant to 2.99 in schizo-dominant subgroups of schizoaffective disorder (the corresponding ratios from Angst et al. 1979b, were 0.59 and 1.02).

Other familial analyses of schizoaffective subtypes have since been undertaken by Baron et al. (1982) and by Kendler et al. (1986). In the study by Baron et al. (1982), the morbid risk of schizophrenia was 4.1% in relatives of mainly schizophrenic schizoaffectives vs. 0% in relatives of mainly affective schizoaffectives, and the reverse order was found for unipolar disorder in relatives (10.9% morbid risk in the mainly schizophrenic subgroup vs. 26.5% in the mainly affective subgroup). The same pattern was apparent in the analysis by Kendler et al. (1986) of the Iowa 500/Non500 data. In particular, the morbid risk of schizophrenia was 8.2% in relatives of mainly schizophrenic schizoaffectives vs. 3.8% in relatives of other or mainly affective schizoaffectives, and the reverse order was found for affective disorders in relatives (7.3% morbid risk in the mainly schizophrenic subgroup vs. 14.5% in the other or mainly affective subgroup). Kendler et al. (1986) concluded that mainly schizophrenic schizoaffective disorder (including in particular bipolar patients with persistent psychosis) has a close familial relationship to schizophrenia "with no apparent relationship to affective illness" (p. 1003). The authors further hypothesized that the remaining schizoaffective patients are "etiologically heterogeneous, with possible relationships to schizophrenia, affective illness, and remitting psychotic syndromes" (p. 1003).

Although these findings have helped to clarify from a familial perspective the status of "cases in between" schizophrenia and affective disorder, they cannot be considered decisive in the debate concerning the continuum vs. Kraepelinian positions. On the one hand, there is evidence that major subgroups of schizoaffective patients bear a close family relationship to schizophrenic patients, and this is consistent with the Kraepelinian position, whereby one would expect to be able to sort all "cases in between" into schizophrenic and affective familial groups (within the bounds imposed by random variability and misclassification). On the other hand, not all studies have found the same results, and in any case there appears to be a substantial subgroup of schizoaffective patients who cannot be grouped with schizophrenia or affective disorder this time. Furthermore, there is some evidence for a gradient of risk (e.g., the results of Angst and Scharfetter 1990, discussed above) which is at least consistent with a continuum hypothesis.

As stated above, part of the difficulty in addressing this issue is that most studies have been framed in categorical terms consistent with the major diagnostic systems. As a result, there have been wide differences across studies in how study groups (and subgroups) are defined. Paradoxically, however, the solution is not necessarily to adopt a standard classification system. For example, if a universally accepted nosology were unwittingly to classify as "schizoaffective" two etiologically distinct subgroups of patients, then studies based on that nosology would probably not advance our understanding of how schizoaffective disorder is related to schizophrenia and affective disorder. In view of the unsettled nature of this area of research, I would therefore advocate instead that various alternative approaches be explored whenever possible, in hopes that convergent validity will eventually reveal the correct picture.

An example of a promising alternative approach is that employed by Kendell and Gourlay (1970), who used a discriminant function analysis to describe their patients in dimensional rather than categorical terms. This approach is particularly relevant to the debate between continuum and Kraepelinian models, because it could reveal subtle or complex patterns that perhaps are being obscured by strictly categorical approaches. Of course, in analyses of family data, the necessity of computing morbidity risks for groups of subjects indicates that a hybrid approach based on a combination of dimensional and categorical methods is more practical than a strictly dimensional approach. Previously, our research group took such a hybrid approach in using logistic regression to propose preliminary clinical and research criteria for schizoaffective subtypes (Tsuang et al. 1986), to examine long-term outcome in three subtypes of schizoaffective disorder (Tsuang 1990), and to investigate morbidity risks of schizophrenia and affective disorders in the undifferentiated subtype (Tsuang 1991).

To illustrate the application of this approach to the topic at hand, we report here the preliminary family study results for the heterogeneous group of 271 broadly defined "schizoaffective" patients from the Iowa Non500 Study,

Table 1. Morbidity risk of schizophrenia in relatives of patients with three subtypes of schizoaffective disorders compared with relatives of patients with major psychoses

Proband group	Age adj. no. relatives (BZ)[a]	Relatives with schizophrenia	Morbidity risk (%)
Schizoaffective subtype			
Schizophrenic (SA-S)	254.0	9	3.5
Undifferentiated (SA-U)	152.5	10	6.6
Affective (SA-A)	255.5	11	4.3
Major psychoses			
Schizophrenia (S)	362.0	20	5.5
Affective disorder (AD)	679.0	15	2.2

Comparison of morbidity risks: SA-S:S; SA-U:S; SA-A:S
SA-S:AD; SA-U:AD*; SA-A:AD

*$p < .05$
[a] BZ (Bezugsziffer), age-adjusted size of the sample based on a risk period of 15–39 years of age for schizophrenia.

Table 2. Morbidity risk of affective disorder in relatives of patients with three subtypes of schizoaffective disorder compared with relatives of patients with major psychoses

Proband group	Age adj. no. relatives (BZ)[a]	Relatives with affective disorder	Morbidity risk (%)
Schizoaffective subtype			
Schizophrenic (SA-S)	202.5	19	9.4
Undifferentiated (SA-U)	123.5	16	13.0
Affective (SA-A)	203.0	19	9.4
Major psychoses			
Schizophrenia (S)	286.5	29	10.1
Affective disorder (AD)	531.0	96	18.1

Comparison of morbidity risks: SA-S:S; SA-U:S; SA-A:S
SA-S:AD**; SA-U:AD; SA-A:AD**

**$p < .01$
[a] BZ (Bezugsziffer), age-adjusted size of the sample based on a risk period of 15–59 years of age for affective disorder.

who were subtyped into more homogeneous subgroupings according to their clinical characteristics, independent of their family data, using the same logistic regression approach employed by Tsuang et al. (1986) and by Tsuang (1990, 1991). Details of the logistic regression analysis are found in Tsuang (1991).

Results are given in Tables 1 and 2 for comparisons of three subgroups with schizophrenic and affective disorder probands from the Iowa 500 Study (Tsuang 1991). Based on their clinical characteristics, these subgroups were labeled "schizoaffective, schizophrenic subtype" (SA-S), "schizoaffective,

affective subtype" (SA-A), and "undifferentiated" (SA-U). The morbidity risks of major psychoses in the relatives of these three subgroups were then compared with the morbidity risks of major psychoses in the relatives of Iowa 500 Study probands with schizophrenia and affective disorders.

Table 1 gives the comparisons involving the morbidity risk of schizophrenia. Contrary to expectation based on their clinical characteristics, the schizophrenic subtype had the lowest morbidity risk (among subgroups) of schizophrenia and the undifferentiated subtype had the highest risk. In particular, there was no evidence of a gradient of morbid risk of schizophrenia increasing from SA-A to SA-U, SA-S, and finally schizophrenia. Furthermore, differences between subgroups were not statistically significant.

In the comparisons involving the morbid risk of affective disorder in relatives (Table 2), there was an increase in risk from SA-S to SA-U to affective disorder, but SA-A did not fit into this gradient and instead had the same morbid risk of affective disorder as SA-S, contrary to expectation.

These illustrative results therefore cannot be said to be consistent with either the continuum or the Kraepelinian models. On the one hand, the findings, although based on a dimensional analytic approach, do not fit neatly into a continuum framework, which would lead one to expect a gradient in the relative morbid risk of schizophrenia compared with affective disorder (Crow 1990a,b). Whereas in Angst and Scharfetter (1990) the schizophrenia: affective disorder morbidity risk ratio jumped from 0.92 in affect-dominant schizoaffective disorder to 2.99 in schizo-dominant schizoaffective disorder, the corresponding ratios in our data ranged from 0.46 (SA-A) to 0.51 (SA-U) to 0.37 (SA-S). On the other hand, these analyses have not provided us with a clear basis for allocating "cases in between" unequivocally to either schizophrenic or affective disorder prototypes, and therefore agree with the conclusion of Angst et al. (1979b) that the findings, though crucial to the resolution of the debate concerning the nature of psychosis, are currently ambiguous.

Inasmuch as we view these results as preliminary, it is worth considering the limitations of the dimensional approach illustrated here. One possibility is that the results are not representative because they are distorted by random variability. This is somewhat unlikely, in view of the respectable sample sizes employed, but the possibility exists and highlights the need for systematic replication in independent samples. It is also possible that the pattern of results has been distorted by more systematic factors. For example, the inclusion of gender and associated demographic characteristics (age at onset and marital status) in the logistic regression equation could result in schizoaffective subgroups with clinical characteristics, and possibly morbid risks, that differ substantially from the subgroups that would be obtained by stratifying the analysis by gender. It should be pointed out, however, that Angst et al. (1979b) subdivided schizoaffective probands by gender but found no systematic differences in morbidity risks for male and female subjects.

Another factor with potential nosologic importance is the polarity of the affective component. In an analysis of the Iowa 500/Non500 data using DSM-III, Kendler et al. (1986) found that the pattern of morbidity risks in schizoaffective patients with no history of bipolar features (i.e., depressive subtype) more closely resembled the pattern in schizophrenia, compared with schizoaffective patients with a bipolar history. Other investigators have also found nosologically relevant differences between bipolar and unipolar subtypes (see Levinson and Levitt 1989). A particularly interesting recent example is provided by Maier et al. (1994), who found substantial differences in morbidity risks for bipolar and unipolar subtypes of schizoaffective disorder in a large sample study and concluded that polarity is of fundamental importance in understanding psychotic illness.

Finally, we should consider the possibility that the preliminary results presented here are indeed representative, and that in this border region between schizophrenia and affective disorder empirical relationships are not simple, but complex. For example, Maier et al. (1994) concluded from a controlled family study of schizoaffective disorder that the Kraepelinian dichotomy can be confirmed for schizophrenia compared with bipolar affective disorder, but not in comparison to unipolar depression. As a related but somewhat different example, Kendler et al. (1993) concluded from the case register-based Roscommon (Ireland) Family Study that DSM-III-R schizoaffective disorder "has a close family relationship with schizophrenia but also shares familial factors with affective illness and particularly bipolar affective illness." These examples highlight the need for additional empirical investigation in this area, in particular the need to replicate all major findings in independent samples so that theoretical conclusions are not drawn prematurely, but rather are founded on a solid empirical base.

In conclusion, this illustrative application of a hybrid dimensional/categorical approach to subtyping "cases in between" schizophrenia and affective disorder using family data has not clearly delineated the boundary region between the Kraepelinian categories of psychosis. Family data do indicate that the boundary between schizophrenia and affective disorder consists of schizoaffective disorder. However, the precise lines of separation on either side of schizoaffective disorder, one side with schizophrenia and the other with affective disorder, remain very fuzzy. If the Kraepelinian binary model for major psychoses applied, we would expect that the subtypes of schizoaffective disorder could be combined unambiguously with either schizophrenia or affective disorder, leaving only two distinct categories. If the continuum model held, we would expect a steady gradient in the morbid risk of schizophrenia relative to affective disorder, consistent with the predominance of affective or schizophrenic features. Since neither of these expectations was confirmed, at this point we cannot conclude that schizophrenia is on a psychotic continuum or is a distinct entity.

References

Angst J, Felder W, Lohmeyer B (1979a) Schizoaffective disorders: results of a genetic investigation, I. J Affective Disord 1:139–153

Angst J, Felder W, Lohmeyer B (1979b) Are schizoaffective psychoses heterogeneous? Results of a genetic investigation, II. J Affective Disord 1:155–165
Angst J, Scharfetter C (1990) Schizoaffective Psychosen. Ein Nosologisches Aergernis. In: Lungershausen E, Kaschka WP, Witkowski RJ (eds) Affektive Psychosen. Schattauer, Stuttgart
Baron M, Gruen R, Asnis L, et al (1982) Schizoaffective illness, schizophrenia and affective disorders: Morbidity risk and genetic transmission. Acta Psychiatr Scand 65:253–262
Crow TJ (1990a) Nature of the genetic contribution to psychotic illness – a continuum viewpoint. Acta Psychiatr Scand 81:401–408
Crow TJ (1990b) The continuum of psychosis and its genetic origins. Br J Psychiatry 1990b, 156:788–797
Crow TJ (1991) The search for the psychosis gene. Br J Psychiatry 158:611–614
Feighner JP, Robins E, Guze SB, et al (1972) Diagnostic criteria for use in psychiatric research. Arch Gen Psychiatry 26:57–63
Gershon ES, DeLisi LE, Hamovit J, et al (1988) A controlled family study of chronic psychoses: Schizophrenia and schizoaffective disorder. Arch Gen Psychiatry 45: 328–336
Guze SB, Cloninger CR, Martin RL, et al (1983) A follow-up and family study of schizophrenia. Arch Gen Psychiatry 40:1273–1276
Kendell RE, Gourlay J (1970) The clinical distinction between the affective psychoses and schizophrenia. Br J Psychiatry 134:261–266
Kendler KS, Gruenberg AM, Tsuang MT (1986) A DSM-III family study of the nonschizophrenic psychotic disorders. Am J Psychiatry 143:1098–1105
Kendler KS, McGuire M, Gruenberg AM, et al (1993) The Roscommon Family Study. II. The risk of nonschizophrenic nonaffective psychoses in relatives. Arch Gen Psychiatry 50:645–652
Levinson DF, Levitt ME (1987) Schizoaffective mania reconsidered. Am J Psychiatry 144:415–425
Maier W, Lichtermann D, Minges J, et al (1994) Continuity and discontinuity of affective disorders and schizophrenia: Results of a controlled family study. Arch Gen Psychiatry (in press)
Mendlewicz J, Linkowsky P, Wilmotte J (1980) Relationship between schizoaffective illness and affective disorders or schizophrenia: morbidity risk and genetic transmission. J Affective Disord 2:289–302
Tsuang MT (1979) "Schizoaffective disorder": dead or alive? Arch Gen Psychiatry 36:633–634
Tsuang MT (1990) Follow-up studies of schizoaffective disorders: a comparison with affective disorders. In: Marneros A, Tsuang MT (eds) Affective and schizoaffective psychoses: similarites and differences. Springer, Berlin Heidelberg New York
Tsuang MT (1991) Morbidity risks of schizophrenia and affective disorders among first-degree relatives of patients with schizoaffective disorders. Br J Psychiatry 158: 165–170
Tsuang MT, Winokur G, Crowe RR (1980) Morbidity risks of schizophrenia and affective disorders among first-degree relatives of patients with schizophrenia, mania, depression and surgical conditions. Br J Psychiatry 137:497–504
Tsuang MT, Simpson JC, Fleming JA (1986) Diagnostic criteria for subtyping schizoaffective disorder. In: Marneros A, Tsuang MT (eds) Schizoaffective psychoses. Springer, Berlin Heidelberg New York

Etiology of Schizophrenia: Perspectives from Childhood Psychoses

H. Remschmidt, M. Martin, E. Schulz, and G.-E. Trott

Introduction and Historical Remarks

Schizophrenic psychoses in childhood were delineated as specific psychotic disorders only in the late 1930s (Lutz 1937/1938). But the history of childhood schizophrenia starts earlier. Hermann Emminghaus, who wrote the first textbook of child psychiatry in 1887 with the title *Psychic Disturbances of Childhood*, describes childhood psychoses as "cerebral neurasthenia" and defines these disorders as "neuroses of the brain characterized by a reduction of cognitive abilities, mood changes, sleep disturbances and manyfold anomalies of innervation with a subacute or chronic course and different states of outcome" (Emminghaus 1887, p. 134). Emminghaus was also the first who introduced the developmental perspective into child psychiatry with special focus on psychoses. After Emil Kraepelin, at the end of the past century, first described dementia praecox (Kraepelin 1886) and Eugen Bleuler (1911) introduced the term "schizophrenia", August Homburger stated, in his famous textbook (Homburger 1926), that *childhood schizophrenia* really exists and may be characterized by the following features: (a) A predominant change of the personality of the children in the direction of withdrawal, negativism, and strange and unexpected behavior. We would say today "negative symptoms"; (b) rare delusions, especially in young children; (c) at least two types of manifestation: a slow retarded, hebephrenic form with cognitive deterioration and an acute catatonic form; and (d) premorbid characteristics (e.g., primordial mental retardation, primordial emotional disturbances). Leo Kanner (1943) and Hans Asperger (1944) delineated the meanwhile well-known autistic syndromes (early infantile autism and autistic personality disorder) out of the pool of childhood psychoses. Finally, Karl Leonhard's work has to be mentioned. Leonhard (1986) stated that the group of schizophrenic psychoses comprises very heterogeneous disorders, and to find a unique etiology of schizophrenia is therefore an absolute illusion. Unlike many other authors, he subdivides the schizophrenic disorders into *unsystematic* and *systematic* schizophrenias. According to Leonhard, *unsystematic* schizophrenias are characterized by primarily affective symptoms (e.g., extreme anxiety states, delusions, hallucinations, ideas of reference); their course is acute, sometimes also periodic with a

---

Department of Child and Adolescent Psychiatry, Philipps University, Hans-Sachs-Str. 6, 35033 Marburg, Germany

good remission. They have more in common with affective psychoses than with the group of systematic schizophrenias. *Systematic schizophrenias* show predominantly cognitive disturbances and disturbances of voluntary functions. Their course is predominantly chronic without recovery to the former cognitive level, and the prognosis is poor. As a special form of childhood schizophrenia, Leonhard distinguishes *early infantile catatonia*. He described this form in a very detailed way in the last edition (1986) of his well-known book. The clinical picture of infantile catatonia is characterized by motor symptoms, absence of language development or a very poor language, circumscribed intellectual impairments, negativism, sometimes periodic course, and dominance of the male sex.

Epidemiology and Classification

There are no reliable epidemiological data that could be looked upon as representative. This has to do with different definitions and with the fact that several heterogeneous disorders have been subsumed under the heading "psychoses of childhood". According to Bleuler (1972) and Lutz (1972), about 4% of schizophrenic psychoses begin before the 15th year of life and 0.5–1% before the 10th year of life. With respect to the general population, one can assume that one child among 10 000–15 000 will develop a schizophrenic disorder.

There are, however, not many epidemiological studies that answer questions concerning frequency, age and sex distribution, treatment, and course of childhood and adolescent schizophrenia.

Werry (1992) states that the results of studies carried out earlier than 1975 are questionable, because at that time no internationally accepted classification schemes were available. According to several authors, the following statements can be made: (a) The prevalence of childhood onset schizophrenia (manifestation before 12) is less than one child in 10 000 children between 2 and 12 years (Burd and Kerbeshian 1987). (b) It is estimated that childhood onset schizophrenia is approximately 50 times less frequent than adult onset schizophrenia (Karno and Norquist 1989). (c) Childhood onset schizophrenia is also less frequent than early infantile autism (Burd and Kerbeshian 1987). (d) There is no co-morbid association between autism and schizophrenia, either in childhood or in adult age. The two conditions are not more commonly observed together than would be expected by chance (Volkmar and Cohen 1991). Our own studies show that up to the age of 12, all psychotic states are rare conditions, but that after the 13th year of life there is a remarkable increase of schizophrenia (Remschmidt 1988; Remschmidt and Martin 1992).

Sex Distribution

Another very interesting question is the *sex distribution*. If we look at two samples of ICD-9-diagnosed schizophrenic patients of two clinical centers

(Würzburg and Marburg), we find a nearly equal distribution of the sexes among the total groups of schizophrenic patients from 6 up to 19 years. But if we analyze a subsample of two age-groups, the early-onset group I (with manifestation before 14 years) and group II (first manifestation after 14 years), we find a very clear predominance of the male sex in the younger group. This was demonstrated in two samples (in Würzburg and in Marburg). Table 1 shows two remarkable things: first, the group of very early onset schizophrenia comprises approximately 11% of all schizophrenic patients in childhood and adolescence, and second, the sex relationship in the very young group (group I) is, if we take both studies together, 1:3 in favor of boys. Within the Marburg prospective study, we find a relationship of one girl to 4.5 boys. Though both samples derive from clinical populations, we assume that the data express a realistic relationship, because both settings are the only ones in the regions specialized in these disorders, and one can assume that nearly all schizophrenic patients would be admitted to these centers.

Age and Developmental Stage

At the end of the 1950s and 1960s, it became evident that age and developmental stage may be looked upon as important criteria for the classification of childhood psychoses (GAP 1966; Stutte 1969). This was also demonstrated by several empirical studies (Rutter et al. 1967; Kolvin et al. 1971a–d). Finally, these studies confirmed the notion of Leo Kanner (1943, 1957), who subdivided childhood psychoses into three groups: *early infantile autism*, *childhood schizophrenia*, and *disintegrative psychoses* of childhood, the latter comprising disorders such as dementia infantilis Heller and psychoses related to different kinds of brain damage. These aspects have also influenced the multiaxial classification scheme on the basis of ICD-9 (Rutter et al. 1976) and DSM-III and DSM-III-R of the American Psychiatric Association (APA 1980, 1987). Both classification systems differentiate between early infantile autism, childhood schizophrenia, and early childhood dementia. Most researchers agree that with respect to age and developmental stage, at least four groups of psychoses in childhood and adolescence can be differentiated (Remschmidt et al. 1991).

The *first group* of psychoses comprises different psychotic syndromes, all of which have a slow beginning and a chronic course and manifest themselves before the third year of life. Except for early infantile catatonia, they have no connections with schizophrenia.

The *second group* includes different psychotic states, most of them with acute beginning and different regressive behaviors that manifest themselves between the third and fifth year of life. The connection with schizophrenia is questionable, again with the exception of early infantile catatonia, described by Leonhard (1986). According to him, the connection to schizophrenia is likely.

The *third group* of psychoses are the so-called late-onset psychoses of late childhood to prepuberty, which have a fluctuating or subacute course and a

Table 1. Sex distribution of schizophrenia in two groups of patients with manifestation of the disorder before (group I) and after 14 years of age (group II). All patients were diagnosed according to the ICD-g multiaxial classification and were treated as inpatients. (*WU* Wurzburg, *MR* Marburg)

Study	WÜ: Clinical study (1985–1990)		MR: Prospective study (1991/1992)		Both studies		Both studies
Group	Male	Female	Male	Female	Male	Female	Age-groupe
Group I <14 years	6 (66.7%)	3 (33.3%)	9 (81.9%)	2 (18.1%)	15 (75%)	5 (25%)	20 (11%)
Group II >14 years	55 (49.1%)	57 (50.9)	23 (46%)	27 (54%)	78 (48.1%)	84 (51.9%)	162 (89%)
	61 (50.4%)	60 (49.6%)	32 (52.5%)	29 (47.5%)	93 (51.1%)	89 (48.9%)	182 (100%)

clear relationship to schizophrenia of adolescence and adulthood. A good example here is prepubertal schizophrenia (Stutte 1969; Eggers 1973).

The *fourth group*, finally, is adolescent schizophrenia with manifestation at puberty and adolescence and a very clear relationship to adult schizophrenia.

In his recent review, Werry (1992) distinguishes between early-onset schizophrenia (EOS), defined as beginning in childhood or adolescence (under 16 or 17), and very early onset schizophrenia (VEOS), to be used when the onset is at less than 13 years of age. The argument for the latter group is that the definition is more precise than the term "prepubertal", because puberty may vary in relation to age and most studies that have used the term prepubertal schizophrenia have not taken into consideration the pubertal stages. Werry states further that a review of studies of childhood schizophrenia is complicated by the fact that up to the appearance of ICD-9 and DSM-III, all psychotic disorders of childhood were aggregated into a single category, "childhood schizophrenia", so in many papers it is impossible to differentiate between early infantile autism, childhood schizophrenia, and other psychoses.

Etiological Aspects

As far as etiology of child and adolescent schizophrenia is concerned, several factors have been investigated, with special attention paid to the effect of risk factors and precursor symptoms that may be different in nature and that may either be genetically transmitted or have environmental causes.

Neurointegrative Deficits

Barbara Fish stated in 1975 that a schizophrenic genotype may be manifested in infants by a neurointegrative defect which she called pandysmaturation (PDM) (Fish 1977). Recently, Fish et al. (1992) reviewed the literature with respect to neurointegrative deficits as risk factors for childhood schizophrenia and found that 12 studies replicated a delayed development in infants of schizophrenics and in preschizophrenics. They further replicated the contribution of PDM as a marker for a schizophrenic genotype in a sample of the Jerusalem Infant Development Study (JIDS), following earlier results of this study reported by Markus (1974), who described cerebral dysfunction in offspring of schizophrenics as a possible genetic factor.

Evidence for the hypothesis of a neurointegrative deficit comes also from studies with a focus on visual information processing in schizophrenic children, including evoked potentials (Asarnow et al. 1986), and from adult psychiatry. Castle and Murray (1991), for instance, interpret also the sex differences in schizophrenia in terms of deviant neurodevelopmental processes.

Attentional Dysfunction

For many years, attentional dysfunction has also been looked upon as a biological marker of schizophrenia. According to Holzman (1983), there are four lines of evidence for this association: (a) The replicated fact of attentional dysfunction in schizophrenics, both in episode and remission, (b) attentional deficits in unaffected first-degree adult relatives of schizophrenic patients, (c) attentional deficits as precursor symptoms of schizophrenia in high-risk individuals that predate clinical symptoms by many years (This notion results from high-risk studies), and (d) the hypothesis that attention per se might be genetically transmitted (Cornblatt et al. 1988).

These results raise the question of whether attentional dysfunctions are associated with social dysfunctions in schizophrenia and whether they can also predict social deficits in unaffected adults at risk for schizophrenia. These questions were investigated in a recent study by Cornblatt et al. (1992) in a subsample of the New York High-Risk Project (NYHRP). The main result was that the subjects at risk for schizophrenia were significantly more impaired with respect to social deficits than the other two groups, there being, no difference between the subjects at risk for affective disorders and the normal controls.

Cornblatt et al. (1992) finally proposed a hypothetical model of the role of attentional deficits in the development of schizophrenia, which postulates that the active avoidance of contact with other people resulting from an inability to process interpersonal information due to attentional dysfunction may be a way of reducing stress and thus a method of symptom control.

The results of Asarnow et al. (1986), who found that controlled attentional processes are impaired in schizophrenic children, while more automatic ones are relatively intact, fit into this pattern.

Communication Deficits

Several studies of adult schizophrenia and adult-onset schizotypical personality disorders have provided evidence for a relationship between these two disorders (Lenzenweger and Loranger 1989; Siever et al. 1990). On the other hand, both types of disorders show also a reduced effectiveness in communication (Martin and Chapman 1982). These results raise the question whether the same applies to childhood schizophrenia and childhood schizotypical personality disorder. As the DSM-III-R diagnostic criteria include symptoms similar to those for childhood schizophrenia, an investigation of both groups with respect to communication deficits seems to be reasonable. Caplan et al. (1992) examined 31 schizophrenic children and a control group matched according to sex and mental age with regard to communication deficits and formal thought disorders. The study also aimed to find out if the discourse deficits of schizophrenic children were related to clinical measures of formal thought disorder. Speech samples were collected from the children with the Kiddie Formal Thought Disorders Story Game (Caplan et al. 1989), which was videotaped and rated for formal thought disorder and was also transcribed for coding cohesion and reference patterns.

The results showed similarities to and differences from the communication deficits known from adult schizophrenics. Several of the children's deficits were similar to those found in schizophrenic adults: they speak less than normal subjects of the same mental age; their speech has not enough links to previous utterances (cohesive ties) and not enough references to people, objects, or events that were mentioned in earlier utterances (referential cohesion). Schizophrenic children also often break with slow-off speech in order to refer to objects, people, or events in their immediate surroundings. Those schizophrenic children with loose associations confuse the listener also by the unclear and ambiguous way they refer to their surroundings. With regard to these discourse similarities, the communication deficits of child-onset schizophrenia and those of adult schizophrenia can be compared.

Premorbid Characteristics

In the Marburg prospective study, we included 61 schizophrenic patients who were treated as inpatients at the Marburg University Hospital or in a rehabilitation project at the same hospital during 1991 and 1992. This sample comprised 32 boys and 29 girls of the same age-group. All patients had been classified according to the ICD-9 multiaxial classification scheme. The patients were subdivided into two groups (group I: early onset, under 14 years; group II: first manifestation of schizophrenia beyond 14 years of age). The diagnoses of both subgroups according to DSM-III-R criteria are demonstrated in Fig. 1. As the figure shows, the three major diagnoses are: paranoid schizophrenia (295.3), disorganized type (295.1), and schizoaffective disorder (295.7).

In order to obtain information about positive and negative symptoms and other precursors of schizophrenia, we administered a semistructured interview (IRAOS) which allowed us to check retrospectively schizophrenia-typic symptoms and other symptoms *before* the first manifestation of schizophrenia. This instrument was developed by Häfner and his group (Häfner et al. 1990) and was modified for the special purposes of investigating children and adolescents and their parents by our group. The IRAOS instrument was administered by experienced child psychiatrists asking the parents and using all information that could be obtained from the records. The inter-rater reliability of the instrument was found to be satisfactory, with Kappa values varying between 0.62 and 1.0 (Häfner et al. 1992a,b).

Figure 2 demonstrates the negative and positive symptoms before the manifestation of schizophrenia for the two groups. There is at least a trend, in that group I (very early onset, VEO) shows a longer "precursor time" of negative symptoms than group II.

The same tendency is demonstrated in Fig. 3, which shows the cumulative frequency of positive and negative symptoms before the index admission. Many patients in the total sample ($n = 61$) show both negative and positive

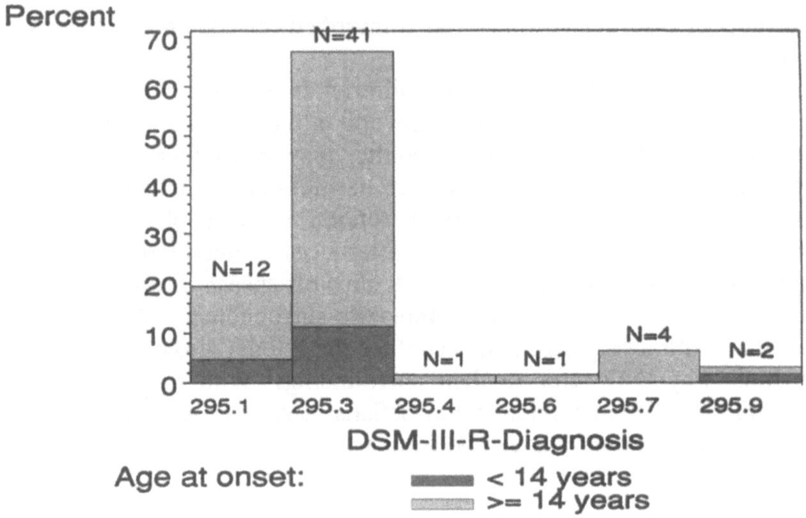

Fig. 1. Distribution of DSM-III-R diagnoses of 61 schizophrenic children and adolescents subdivided into two groups, with first manifestation of the schizophrenic disorder before and after 14 years of age

symptoms before the index admission. At the same time, both categories of symptoms are convergent at the time of the index admission.

If we look at both subgroups (group I and II), more patients of group I show positive and negative symptoms before the index admission than the patients of group II. This is demonstrated in Fig. 4a and b.

These investigations underline the notion of earlier studies that age and developmental stage are decisive variables for the symptomatology of schizophrenic states in childhood and adolescence. This seems to be true at least for the group of very early onset.

In order to get an impression of which symptoms may exist in the premorbid stage of schizophrenia, we developed a checklist that included a catalogue of symptoms which could be classified according to the categories "introversive" and "extraversive symptoms". Examples of symptoms that are characteristic for introversion are: mutism, mental slowness, social isolation, general anxieties, specific anxieties, and obsessive-compulsive symptoms. The dimension extraversion comprised such items as: hypermotoric, dissocial behavior, aggression, and school refusal. According to this categorization of symptomatology, there was no significant difference between the two groups. There was only a nonsignificant trend in the direction of more introversive symptoms in group I.

Another grouping of our symptom checklist was carried out according to four sets of premorbid disturbances, namely developmental disorders, conduct disorders, learning disabilities, and emotional disorders. In all cases, we tried to get the information about premorbid behavior in two ways: through a careful analysis of the case histories and by questioning the parents.

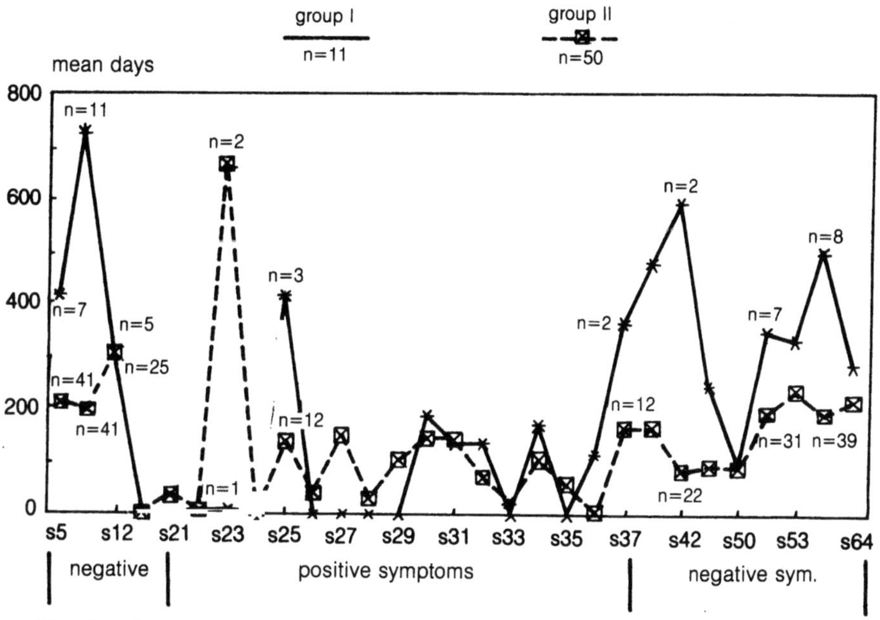

Negative Symptom

5 poor concentration
9 social withdrawal
12 anergia and retardation
13 loss of libido
41 increased distractability
42 disturbance of affect
44 disturbance of speech
50 self-care
52 underactivity
53 slowness
54 social withdrawal
64 loss of interest

Positive Symptoms

21 thought disorder
22 thought insertion
23 thought broadcast
24 thought echo
25 thought block and withdrawal
26 auditory hallucinations
27 verbal hallucinations
28 visual hallucinations
29 other hallucinations
30 delusions of control
31 delusions of reference
32 delusions of persecution
33 expansive delusions
34 delusions of influence
35 primary delusions
36 delusions concerning appearance
37 other delusions

Fig. 2. Negative and positive symptoms in two subgroups of schizophrenic patients with manifestation before 14 (*group I*) and after 14 (*group II*). Symptoms were investigated retrospectively by administering a semistructured interview for the assessment of onset of schizophrenia (IRAOS)

With respect to this subdivision of symptomatology, we found no remarkable differences between the two groups (group I and II). There was a general trend in the direction that developmental disorders appear first, followed by conduct disorders, emotional disorders, and learning disabilities.

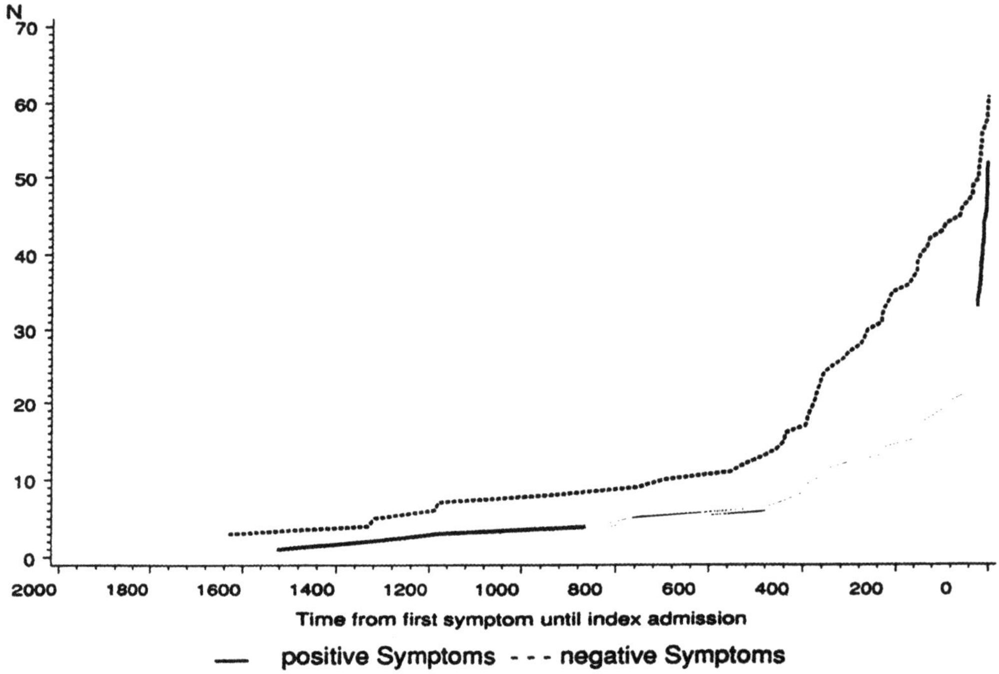

Fig. 3. Retrospective assessment of positive and negative symptoms in a group of schizophrenic children and adolescents (n = 61) by a semistructured interview for the retrospective assessment of the onset of schizophrenia (IRAOS) for up to 2000 days. The first admission (index admission) is marked by 0

Fig. 4. a Positive symptoms and age at onset in two groups of schizophrenic patients with first manifestation of the schizophrenic disorder before 14 (group I, n = 11) and after 14 (group II, n = 50). The first admission (index admission) is marked by 0. The symptoms were assessed retrospectively by the semistructured interview IRAOS. **b** Negative symptoms and age at onset in two groups of schizophrenic patients with first manifestation of the schizophrenic disorder before 14 (group I, n = 11) and after 14 (group II, n = 50). The first admission (index admission) is marked by 0. The symptoms were assessed by an interview with the parents and also by using all information in the records

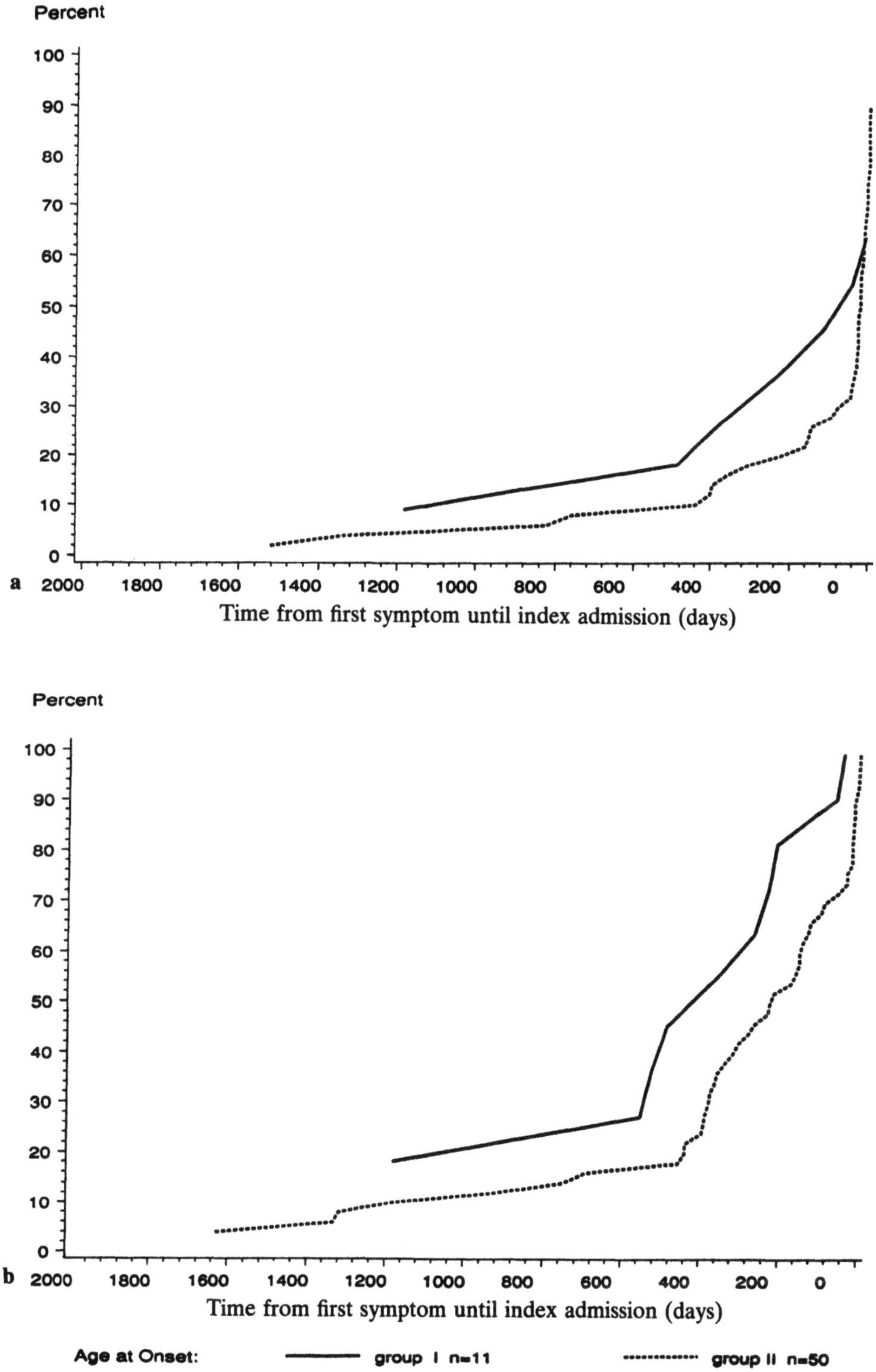
Percent
100
90
80
70
60
50
40
30
20
10
0
a
2000 1800 1600 1400 1200 1000 800 600 400 200 0
Time from first symptom until index admission (days)
Percent
100
90
80
70
60
50
40
30
20
10
0
b
2000 1800 1600 1400 1200 1000 800 600 400 200 0
Time from first symptom until index admission (days)
Age at Onset:
group I n=11
group II n=50

Therapy and Rehabilitation

With regard to *therapy* of schizophrenic psychoses in children and adolescents, five different aspects have to be integrated:

1. Pharmacological treatment of acute psychotic states
2. Pharmacological prevention of relapses
3. Psychotherapeutic measures
4. Family-oriented measures
5. Specific measures of rehabilitation

As we have described the therapeutic programs for schizophrenic children and adolescents elsewhere (Remschmidt 1993; Remschmidt and Martin 1992), we will only delineate here our trials in therapy-refractory schizophrenia with clozapine. The first study carried out by Siefen and Remschmidt in 1986 comprised 21 patients, 12 of whom were below 18 years of age. The indications for administering clozapine were insufficient response to other neuroleptic agents, likelihood of psychotic symptoms becoming chronic, and extensive extrapyramidal side effects from the neuroleptic agents. The patients had had an average of 2.4 hospitalizations and had received on average 2.8 different antipsychotics without adequate therapeutic response. Clozapine was administered over an average of 133 days with an average maximum dose of 415 mg/day (range: 225–800 mg/day) and an average maintainance dose of 363 mg/day (range: 150–800 mg/day). Marked improvement or complete disappearance of most of the remaining psychopathological symptoms was seen in 11 (52%) of 21 subjects, with at least some improvement seen in an additional six (29%). Vegetative side effects such as daytime weariness, dizziness, orthostatic hypotonia, and hypersalivation were usually transient. Deviations were of no clinical significance and not more frequent than with other neuroleptic agents, and the values returned to normal when treatment was continued. The authors conclude that, all in all, clozapine is a useful addition to the group of neuroleptic drugs available for the treatment of adolescent schizophrenia.

In a further study, Remschmidt et al. (1992) reported on a retrospective and a prospective study with clozapine in adolescent schizophrenia. Related to these investigations, a study sample of 41 adolescent patients with ICD-10 diagnoses of schizophrenia (mean age: 18.2 years, range: 13–22 years) was treated with the atypical neuroleptic clozapine (mean dosage: 330 mg/day, range: 50–800 mg/day).

These 41 patients received the atypical neuroleptic clozapine after having fulfilled the following conditions:

1. Nonresponse under conventional neuroleptic medication, including at least two neuroleptic drugs – usually haloperidol and fluphenazine in recommended dosage. This applied to 35 patients (85.4%) of the clozapine group.
2. Deterioration of the symptomatology under pretreatment with typical neuroleptics. This applied to 19 patients (46.3%).
3. Extreme side effects under conventional neuroleptic treatment, observed in 18 patients (43.9%).

The most common side effects under typical neuroleptic treatment were dyskinesias and akathisia, which occurred in about 52% of these patients. Severe cholinergic side effects with atonia of the bowel and string bladder occurred in about 15% under the pretreatment conditions. Therefore, the indication for clozapine treatment was a result of clinical nonresponse, worsening of symptoms, and severe side effects under conventional treatment strategies. With regard to the risk of granulocytopenia, the clozapine medication was administered after a very careful analysis of the individual case in a consensus conference, and all requirements of the guidelines for clozapine treatment were strictly fulfilled. The 41 patients received clozapine after an average duration of their schizophrenic disorder of 3.3 years.

The following results were obtained: In seven patients (17.1%), clozapine medication had to be interrupted due to severe unwanted side effects. After 2 weeks of 100 mg/day of clozapine in combination with carbamazepine (400 mg/day) one patient developed a severe worsening of symptoms with stupor. After 2 weeks of clozapine (250 mg/day) one patient developed a beginning ileus. In two patients leukopenia (2.9 and 2.5/nl) without agranulocytosis terminated clozapine medication. In two other patients hypertonia, tachycardia, and electrocardiographic abnormalities occurred. In one case liver enzymes (GOT and GPT) rose to ten times normal range values.

Three patients (7.3%) did not respond to clozapine medication. Thus, ten (24.4%) of 41 clozapine-treated schizophrenics did not benefit from the atypical neuroleptic medication.

In 31 (75.6%) of these 41 schizophrenics without any pretreatment response, or with worsening of symptoms and side effects under conventional neuroleptic drugs, a remarkable improvement of schizophrenic symptomatology was observed. The majority of these patients were able to participate in a comprehensive rehabilitation program, and three of them (7.3%) showed a complete remission of schizophrenic symptomatology.

Further, it was evident that type-I schizophrenias were better influenced by clozapine medication than type-II schizophrenias. The rate of improvement of the positive symptomatology was approximately 65%. This applied especially to delusions, hallucinations, positive thinking disorders, excitement, and attention.

Clozapine was not as effective with regard to negative symptomatology. Especially such symptoms as anhedonia, affective flattening, and autistic behavior were not improved significantly. But it was interesting that clozapine was somewhat effective concerning such symptoms as anergia (up to 11%), mute behavior (14%), bizarre behavior (22%), and thought blocking (22%).

The results of the ongoing prospective study, the design of which was described by Remschmidt et al. (1992), are not yet available.

Prognosis and Outcome

In contrast to schizophrenia in adulthood, there are not many studies concerning schizophrenic adolescents that take into account the course of their disease.

Table 2. Classification of the course of schizophrenia according to DSM-III-R criteria in two groups of patients with onset before 14 years (n = 11) and after 14 years (n = 50)

Classification of course according to DSM III-R	Age at onset: Group I (<14 years) N	%	Group II (>14 years) N	%	All N	%
Classification of course						
.00 Unspecified	0	0	5	10	5	8
.01 Subchronic	1	9	10	20	11	18
.02 Chronic	9	81	29	58	38	62
.03 Subchronic with acute exacerbation	0	0	1	2	1	1
.04 Chronic with acute exacerbation	1	9	2	4	3	4
.05 In remission	0	0	3	6	3	4
All	11	100	50	100	61	100

According to Weiner (1982), the prognosis of schizophrenia starting at prepuberty and adolescence is unfavorable compared with schizophrenic disorders with a first manifestation in adulthood. About the same rate of schizophrenic adolescents and adults reach full remission (23% vs. 25%), but only 25% of adolescent schizophrenic subjects achieve a partial remission, compared with about 50% of adult schizophrenic patients. Finally, in 52% of adolescent schizophrenic subjects a chronic course of the disorder is observed, in contrast to only 25% of adult schizophrenic individuals.

From various follow-up studies, the following conclusions can be drawn (Remschmidt 1988):

1. Schizophrenic psychoses with manifestation before the tenth year of age have a poor prognosis (Annell 1963; Eggers 1967). This applies also to schizophrenic psychoses beginning before the age of 14 years (very early onset psychoses), and was demonstrated by our prospective study. The course of both groups was classified according to the criteria of DSM-III-R and ICD-10. This is demonstrated in Tables 2 and 3. Table 2 demonstrates that nine patients of group I show a chronic course, with only one characterized as subchronic and another one as chronic with acute exacerbations. Table 3 (ICD-10 criteria) shows that all 11 patients can be classified according to the continuous type of course. The picture is different concerning group II, but it is still remarkable that 78% of the patients who belong to group II can be classified into the category of subchronic or chronic course according to DSM-III-R criteria, or 56% (continuous course) according to ICD-10 criteria.
2. Patients with acute manifestation of their disorder and with productive schizophrenic syndromes such as hallucinations and delusions (positive

Table 3. Pattern of course according of ICD-10 criteria in two groups of schizophrenic patients: group I, onset before age 14 (n = 11); group II, onset after age 14 (n = 50)

Patterin of course according to ICD 10	Age at onset					
	Group I		Group II		All	
	N	%	N	%	N	%
Pattern of course						
.00 Continuous	11	100	28	56	39	63
.02 Episodic with stable deficit	0	0	1	2	1	1
.03 Episodic remittent	0	0	2	4	2	3
Incomplete remission	0	0	9	18	9	14
.05Complete remission	0	0	1	2	1	1
.09 Period of observation, less than on	0	0	9	18	9	14
All	11	100	50	100	61	100

symptoms) have a better prognosis than those with a slow manifestation and with continuous impairment of cognitive functions and/or depressive states.

3. The premorbid personality is also of great importance. Patients who in the premorbid phase were socially active, intelligent, and well-integrated children and adolescents have a better prognosis than those who were cognitively impaired, shy, introverted, and withdrawn before the manifestation of their schizophrenic disorder (Remschmidt et al. 1988; Martin 1991).
4. Finally, the prognosis is better in patients without any family burden of schizophrenia, in those who have good cooperation with the family, and where there is rapid improvement during inpatient treatment (Weiner 1982).

Discussion

Schizophrenic psychoses in childhood are important but rare disorders within the spectrum of psychoses. There is no doubt that schizophrenia exists also in childhood, but there is a remarkable increase of schizophrenic disorders between the 12th and the 15th year of age. There are several psychotic states of early childhood that have nothing to do with schizophrenia, with the exception of early infantile catatonia, described by Karl Leonhard (1986). There is the general notion that the relationship of childhood schizophrenia to adolescent and adult schizophrenia increases with age and progress in development. Some authors (Stutte 1969; Eggers 1973) described prepubertal schizophrenia, but, according to Werry (1992), the precondition for such a distinction would be to diagnose very clearly the stage of prepubertal or pubertal development, which was not done in most cases. Therefore, Werry (1992) proposes distinguishing between early-onset schizophrenia (under 16 or 17 years of age) and very early onset (below 13 years of age). It has to be taken into consideration that studies done before the appearance of ICD-9 and DSM-III did not use standardized criteria. So the definition of schizophrenia in terms of modern terminology might be questionable.

A very clear differentiation was made in the literature by Leo Kanner (1943, 1957), who subdivided childhood psychoses into three groups: early infantile autism, childhood schizophrenia, and disintegrative psychoses. This subdivision became part of all classification systems. But with the introduction of the ICD-10 system, early infantile autism is no longer categorized under the term psychoses, but rather as a pervasive developmental disorder. This is in line with follow-up studies that have shown very clearly that there is no comorbid association between autism and schizophrenia, either in childhood or in adult age (Volkmar and Cohen 1991).

As far as the classical subgroups of schizophrenia are concerned, they can be distinguished also in childhood and adolescent schizophrenia, but they do not seem to be very important with regard to prognosis and outcome. A better predictor of outcome is the subdivision of positive and negative symptoms (Remschmidt et al. 1991).

As far as the very early manifestation is concerned, a preponderance of the male sex can be observed and a poor prognosis for very young cases, as compared with the manifestation of the disorder during adolescence. In our view, it seems to be important that precursor symptoms of schizophrenia can be observed also in children and adolescents months and years before the first manifestation of the disorder. This is not so astonishing if we take into consideration that several deficits, including attentional dysfunctions, neurointegrative deficits, and communication deficits, have been observed in these children and some of them also in children of schizophrenic parents who, at the time of investigation, were not or not yet schizophrenic (Cornblatt et al. 1992).

With regard to childhood and adolescent schizophrenia, there is no evidence for a psychotic continuum including other psychotic states than schizophrenia, but there are arguments for a continuum of schizophrenic states along the lines of positive and negative symptomatology.

Therapy in childhood and adolescent schizophrenia should always be multidimensional, integrating pharmacological, psychotherapeutic, family-oriented approaches and measures of rehabilitation. A very serious question is the treatment of therapy-refractory schizophrenia, seen in approximately 40% of all patients. With regard to these patients, the atypical neuroleptic clozapine has proven to be useful and effective, influencing positive and negative symptomatology as well.

References

Annell A (1963) The prognosis of psychotic syndromes in children. Acta Psychiatry Scand 39:235–241

Asarnow R, Sherman T, Strandburg R (1986) The search for the psychobiological substrate of childhood onset schizophrenia. J Am Acad Child Psychiatry 26:601–604

Asperger H (1989)Die "Autistischen Psychopathien" im Kindesalter. Arch Psychiatrie Nervenkrankh 117:76–136
Bleuler E (1911) Dementia praecox oder Gruppe der Schizophrenien. In: Aschaffenburg G (ed) Handbuch der Psychiatrie, spezieller Teil, vierte Abteilung. Deuticke, Wien-Leipzig
Bleuler E (1972) Die schizophrenen Geistesstörungen. Thieme, Stuttgart
Burd L, Kerbeshian J (1987) A North Dakota prevalence study of schizophrenia presenting in childhood. J Am Acad Child Adolesc Psychiatry 26:347–350
Caplan R, Guthrie D, Fish B, Tanguay PE, David-Lando G (1989) The Kiddy Formal Thought Disorder Rating Scale (K-FTDS). Clinical assessment, reliability, and validity. J Am Acad Child Adolesc Psychiatry 28:208–216
Caplan R, Guthrie D, Foy JG (1992) Communication deficits and formal thought disorder in schizophrenic children. J Am Acad Child Adolesc Psychiatry 31:151–159
Castle DJ, Murray RN (1991) The neurodevelopmental basis of sex differences in schizophrenia. Psychol Med 21:565–575
Cornblatt WA, Risch N, Faris G (1988) The Continuous Performance Test, Identical Pairs Version (CPT-IP). I. New findings about sustained attention in normal families. Psychiatry Res 26:223–238
Cornblatt BA, Lenzenweger MF, Dworkin RH, Erlenmeyer-Kimling L (1992) Childhood attentional dysfunction predicts social deficits in unaffected adults at risk for schizophrenia. Br J Psychiatry 161 [Suppl 18]:59–64
Eggers C (1967) Prognose und Verlauf kindlicher und präpuberaler Schizophrenien. Med. Dissertation, Marburg
Eggers C (1973) Verlaufsweisen kindlicher und präpuberaler Schizophrenien. Springer, Berlin Heidelberg New York
Emminghaus H (1887) Die psychischen Störungen des Kindesalters. Laupp, Tübingen
Fish B (1977) Neurobiological antecedents of schizophrenia in children: evidence for an inherited, congenital neurointegrative defect. Arch Gen Psychiatry 34:1297–1313
Fish B, Marcus J, Hans SL, Auerbach JG, Purdue S (1992) Infants at risk for schizophrenia: sequelae of genetic neurointegrative defect. Arch Gen Psychiatry 49:221–235
Group for the Advancement of Psychiatry (1966) Psychopathological disorders in childhood: theoretical considerations and a proposed classification. GAP, New York
Häfner H, Riecher H, Maurer K, Meissner S, Schmidtke A, Fätkenheuer B, Löffler W, Anderheiden W (1990) Ein Instrument zur retrospektiven Einschätzung des Erkrankungsbeginns bei Schizophrenie (Instrument for the retrospective assessment of the onset of schizophrenia – "IRAOS") – Entwicklung und erste Ergebnisse. Z Klin Psychologie 19:230–255
Häfner H, Riecher-Rössler A, Hambrecht M, Maurer K, Meissner S, Schmidtke A, Fätkenheuer B, Löffler W, Anderheiden W (1992a) IRAOS: an instrument for the assessment of onset and early course of schizophrenia. Schizophrenia Res 6:209–223
Häfner H, Riecher-Rössler A, Maurer K, Fätkenheuer B, Löffler W (1992b) First onset and early symptomatology of schizophrenia. Eur Arch Psychiatry Clin Neurosci 242:109–11
Holzman PS (1983) Smooth pursuit eye movements in psychopathology. Schizophr Bull 9:33–36
Homburger A (1926) Vorlesungen über Psychopathologie des Kindesalters. Springer, Berlin Heidelberg New York
Kanner L (1943) Autistic disturbances of affective contact. Nerv Child 2:217–250
Kanner L (1957) Child psychiatry, 3rd edn. Blackwell, Oxford
Karno M, Norquist GS (1989) Schizophrenia: epidemiology. In: Kaplan HI, Saddock BJ (eds) Comprehensive textbook of psychiatry, 5th edn, vol I. Williams & Wilkins, Baltimore
Kolvin E, Ounsted C, Humphrey M, McNay A (1971a) Six studies in the childhood psychoses. II. The phenomenology of childhood psychoses. Br J Psychiatry 118:385–395

Kolvin E, Ounsted C, Richardson LM, Garside RF (1971b) Six studies in the childhood psychoses. III. The family and social background in childhood psychoses. Br J Psychiatry 118:396–402

Kolvin E, Garside RF, Kidd JSH (1971c) Six studies in the childhood psychoses. IV. Parental personality and attitude and childhood psychoses. Br J Psychiatry 118:403–406

Kolvin E, Ounsted C, Roth M (1971d) Six studies in the childhood psychoses. V. Cerebral dysfunction and childhood psychoses. Br J Psychiatry 118:407–414

Kraepelin E (1886) Psychiatrie. Ein Lehrbuch für Studierende und Ärzte. Barth, Leipzig

Lenzenweger MF, Loranger AW (1989) Detection of familiar schizophrenia using a psychometric measure of schizotypy. Arch Gen Psychiatry 46:902–907

Leonhard K (1986) Die Aufteilung der endogenen Psychosen und ihre differenzierte Ätiologie. 6. Auflage. Akademie-Verlag, Berlin

Lutz J (1937/1938) Über Schizophrenie im Kindesalter. Schweiz Arch Neurol Neurochir Psychiatr 39:335–372, 40:141–163

Lutz J (1972) Kinderpsychiatrie. Rotapfel, Zürich

Markus J (1974) Cerebral functioning in offspring of schizophrenics: a possible genetic factor. Int J Ment Health 3:57–73

Martin M (1991) Der Verlauf der Schizophrenie im Jugendalter unter Rehabilitationsbedingungen. Enke, Stuttgart

Martin ME, Chapman LJ (1982) Communication effectiveness in psychosis-prone college students. J Abnorm Psychol 91:420–425

Remschmidt H (1988) Schizophrene Psychosen im Kindesalter. In: Kisker KP, Lauter H, Meyer J-E, Müller C, Strömgren E (eds) Psychiatrie der Gegenwart. Bd. 7: Kinder- und Jugendpsychiatrie. Springer, Berlin Heidelberg New York

Remschmidt H (1993) Schizophrenic psychoses in children and adolescents. Triangle 32:15–24

Remschmidt H, Martin M (1992) Die Therapie der Schizophrenie im Jugendalter. Dtsch Arztebl 89:387–396

Remschmidt H, Martin M, Albrecht G, Gerlach G, Rühl D (1988) Der Voraussagewert des Initialbefundes für den mittelfristigen Rehabilitationsverlauf bei jugendlichen Schizophrenen. Nervenarzt 59:471–476

Remschmidt H, Martin M, Schulz E, Gutenbrunner C, Fleischhaker C (1991) The concept of positive and negative schizophrenia in child and adolescent psychiatry. In: Marneros A, Andreasen NC, Tsuang MT (eds) Negative versus positive schizophrenia. Springer, Berlin Heidelberg New York

Remschmidt H, Schulz E, Martin M (1992) Die Behandlung schizophrener Psychosen in der Adoleszenz mit Clozapin (Leponex). In: Naber D, Müller-Spahn F (eds) Clozapin. Pharmakologie und Klinik eines atypischen Neuroleptikums. Eine kritische Bestandsaufnahme. Schattauer, Stuttgart

Remschmidt H, Schulz E, Martin M, Warnke A, Trott G-E (1994) Childhood-onset schizophrenia: history of the concept and recent studies. Schizophrenia Bulletin 20:727–745

Rutter M, Lockyer L, Greenfield D (1967) A five to fifteen-year follow-up study of infantile psychoses. II. Social and behavioural outcome. Br J Psychiatry 113:1189–1199

Rutter ML, Shaffer D, Sturge C (1976) A guide to a multi-axial classification scheme for psychiatric disorders in childhood and adolescence. Frowde, London

Siefen G, Remschmidt H (1986) Behandlung mit Clozapin bei schizophrenen Jugendlichen. Z Kinder Jugendpsychiatr 14:245–257

Siever LJ, Silverman J, Horvat TB (1990) Increased morbid risk for schizophrenia-related disorders in relatives of schizotypal personality disordered patients. Arch Gen Psychiatry 47:634–670

Stutte H (1969) Psychosen des Kindesalters. In: Schmid F, Asperger H (eds) Neurologie – Psychologie – Psychiatrie. (Handbuch der Kinderheilkunde, Bd. VIII/1). Springer, Berlin Heidelberg New York
Volkmar FR, Cohen DJ (1991) Comorbid association of autism and schizophrenia. J Am Acad Child Adolesc Psychiatry 148:1705–1707
Weiner IB (1982) Child and adolescent psychopathology. Wiley, New York
Werry JS (1992) Child and adolescent (early onset) schizophrenia: a review in the light of DSM-III-R. J Autism Devel Disord 22:601–624

Psychotic Continuum or Distinct Entities: Discussion

J. Angst

Marneros exposed very clearly the controversy between "psychotic continuum or distinct entities" with reference to schizophrenia, schizoaffective psychoses, and affective disorders. Is there only one disorder (unitary concept), two (Kraepelin), a multitude (Wernicke, Kleist and Leonhard), or a spectrum without clear borders between the subgroups (Angst and Scharfetter)? In other words, do the subgroups of functional psychoses differ qualitatively or only quantitatively in regard to psychopathology or phenomenology? In order to suggest distinct clinical entities, syndromal stability has first of all to be shown by prospective clinical studies.

In this respect, the typology of Leonhard has not yet been shown to be valid; furthermore, the breeding true of these subtypes within families has never been shown by studies where the investigator was blind to the proband's diagnosis. On the contrary, the studies on the course of endogenous psychoses carried out by Marneros and his group documented convincingly that a syndromal stability was present only in cases lying at the ends of the spectrum. In the center, where schizoaffective states are located, there is high instability on the psychopathological level, and this is also true for subtypes of schizophrenia. Finally, there is no bimodal, but a unimodal distribution of frequencies of phenomenology or course characteristics between schizophrenia and affective disorders. The schizoaffectives are the true bridge between them.

The presentation of Mundt is in agreement with these findings. Especially the concept of Janzarik postulates the schizoaffective group as the true center of the spectrum, based on his structural dynamic psychopathology.

Mundt pointed out correctly that no pathognomonic symptoms of schizophrenia exist, a fact which was always stressed by Manfred Bleuler. This is so for Schneiderian first-rank symptoms, basic symptoms, negative symptoms, and word associations. Mundt suggests a transnosological pychopathology as an alternative, which is a fruitful hypothesis for further research.

Mundt's paper is also interesting from a methodological point of view, because it illustrates the importance of studies on the whole spectrum of functional syndromes in a standardized way, in order to examine its elements. All sorts of analyses (cluster, factor, or discriminant) require large representative patient samples, and the studies have to be repeated in order to achieve reproducible sample-independent, generalizable results.

Psychiatrische Universitätsklinik Zürich, Postfach 68, 8029 Zürich 8, Switzerland

A special problem mentioned by Mundt is the controversy between the hierarchical and the spectrum approach. From a physiological point of view, we have to assume the existence of diverse functional brain subsystems, which are orgnized and partially coordinated in hierarchies. Therefore, pathological processes may involve one, two, or several subsystems to a variable extent, creating a multitude of psychopathological syndrome profiles. Thus their polymorphism does not yet prove underlying qualitatively different etiologies; the same process could involve one or several systems.

Tsuang tried to validate the nosological concepts of schizophrenia, schizoaffective, and affective psychoses with data from his family studies, including the non-Iowa 500, which means cases which did not qualify for the diagnoses of schizophrenia or affective disorders. In this context one has to be aware that the negatively defined residual group includes not only schizoaffective but also schizophreniform psychoses. The results of this approach are therefore not directly comparable with the findings of Angst and Scharfetter. The two studies differ in the definitions of schizophrenia and schizoaffective psychoses. Schizophreniform psychoses are included under schizophrenia by Angst and Scharfetter, whereas in Tsuang's study they belong to the group of schizoaffectives. This is true for the diagnoses of the probands, but especially for the morbid-risk figures obtained for their relatives. What we really need is a subtyping of the secondary cases among relatives by the same criteria us used for probands, in order to analyze the differences between the two studies. A very narrow concept of schizophrenia, as applied by the American study based on DSM diagnoses, may give an artificially clean picture of a breeding true of schizophrenia.

A more sophisticated approach to family studies consists in comprehensive assessments of the individual profiles of symptoms or syndromes, applying the same instrument to probands and relatives. In this case the studies are not based on diagnoses only but on syndrome profiles. Each item or syndrome must be operationalized and reliable. For this purpose we developed in Zurich a syndrome checklist called SSCL-16 (Angst et al. 1992), which is currently being used in family studies in the United States. The similarity between probands and relatives can then be computed, based on feature vectors of clinical data, applying multidimensional scaling and cluster analysis as described by Angst et al. (1983).

The present state of our knowledge is inconclusive. The hypothesis of a continuum of endogenous psychoses may be too simple, but it is not yet disproved. A more refined hypothesis would include the considerations of Maier's presentation, distinguishing different predispositions, for instance, specific, unspecific, including personality disorders, psychoses (unspecific), and syndromal spectra (manic, depressive, schizophrenic).

Child psychiatric research has not examined the question specifically, and there are no data available which can give us an answer. Until better data are available, we can assume that "we are all correct but we disagree," or, more probably, "we are all wrong and we disagree." This is the usual outcome of such symposia, suggesting more research in the field in order to disagree again in a way at a future meeting.

Part II

The Schizophrenia Spectrum: Perspectives from Neuroimaging*

N. C. Andreasen[1] and M. Flaum[2]

The growth of neuroimaging techniques over the past decade has provided psychiatrists with a set of tools which can be used to pursue some of the most fundamental questions in research in mental illness. The neuroimaging techniques have grown steadily and now include methods that permit us to study brain anatomy in increasingly fine detail, such as computerized tomography (CT) and magnetic resonance (MR), as well as the various functional techniques. The latter include echo planar MR, MR spectroscopy, single photon-emission computed tomography (SPECT), and positron-emission tomography (PET); these functional techniques permit evaluation of brain physiology and chemistry. The goals of investigators who apply neuroimaging research techniques in psychiatry are to study basic brain–behavior relationships, and ultimately to identify the mechanisms that produce mental illnesses. In its current form, neuroimaging research aims to achieve these goals at multiple levels, i.e., to understand illnesses as diseases, as syndromes, or as symptom complexes.

The Complexity of the Schizophrenia Spectrum

Achieving the goal of understanding the underlying mechanisms of various mental illnesses is made considerably more complicated by the complexity of these illnesses at the descriptive level. If we focus on a single category of illness, as is the purpose of this symposium, it soon becomes clear that even with a narrow focus, the purview must necessarily be broad. In the case of the schizophrenia spectrum, we are not clear about how to define the disorder, how to determine its boundaries, or how to identify its characteristic features.

*This research was supported in part by NIMH Grants MH 31593, MH40856 and MHCRC 43271, The Nellie Ball Trust Fund, Iowa State Bank and Trust Company, Trustee, and a Research Scientist Award, MH 00625.
[1] The Department of Psychiatry, The University of Iowa College of Medicine, MHCRC, 2911 JPP, 200 Hawkins Drive, Iowa City, Iowa 52242, USA
[2] The Department of Psychiatry, The University of Iowa College of Medicine, 200 Hawkins Drive, Iowa City, Iowa 55242, USA

Definition of a Schizophrenia Spectrum

In spite of nearly a century of study and investigation, the concept of schizophrenia remains ill-defined. The DSM definition is, at best, a provisional agreement to characterize schizophrenia as a disorder defined by the presence of at least one positive symptom in conjunction with a minimum 6-month duration and some deterioration in functioning. The presence of a full affective syndrome is used as an exclusionary criterion, but (as anyone who works closely with "schizophrenic" patients knows only too well), some have in fact met criteria for a full affective syndrome sometime during the course of the illness, and some are given the diagnosis of affective disorder on their first or second admission before the course and symptoms become clearly characteristic of schizophrenia. Thus, although we have an agreed-upon definition of disorders such as schizophrenia, schizophreniform disorder, schizoaffective disorder, manic disorder, major depressive disorder, or schizotypal disorder, the boundaries between these disorders are not as clear in real life as they are on the black and white pages of diagnostic manuals. Controversy remains as to the importance of a variety of defining aspects: longitudinal course and outcome, the presence of deterioration, the characteristic cross-sectional phenomenology, the significance of negative symptoms, the significance and interrelationship of various positive symptoms, and the relationship of affective syndromes.

These controversies are neither academic nor trivial. Many of the debates in journals or the halls of academia about whether schizophrenics respond to treatment, have identifiable biological correlates, have a disease that can be identified by a single major locus, or have a poor prognosis are at least partly explicable by the variability of the concept across time and space. Kraepelin's definition of schizophrenia (Kraepelin et al. 1919) was relatively narrow and emphasized an early age of onset and a deteriorating course ("dementia praecox"). Bleuler (1950) broadened the concept substantially by introducing the term "schizophrenia," emphasizing cross-sectional phenomenology as more important than outcome, and arguing that the disease should be defined in terms of "fundamental" symptoms such as affective blunting or avolition rather than the positive symptoms of psychosis, which he viewed as "accessory" and nonspecific. The introduction of Schneiderian concepts in Great Britain in the late 1960s and early 1970s shifted the pendulum back to a narrow definition again (Mellor 1970; Wing et al. 1974; Cooper et al. 1972). This narrow definition, with an emphasis on positive and Schneiderian symptoms, was institutionalized in DSM-III, DSM-III-R, and ICD-10, although it is less prominent in DSM-IV. Further, current US definitions of schizophrenia draw sharper boundaries between schizophrenia and affective disorders than are placed elsewhere in the world; in general, these changing and variable concepts make past research difficult to interpret and require that we exercise caution in grouping studies together and generalizing from them by using meta-analysis or the "box score" approach.

These changing concepts also have important implications for the study of major psychoses. They indicate that there is no truly "certain" way to define schizophrenia or affective disorder. For example, the disorder referred to as "schizophrenia" is probably not unlike mental retardation, in that it is very likely a final common pathway or syndrome with variable severity and variable phenomenology, although with a core concept comprised of a mixture of positive and negative symptoms and a degree of conceptual and social disorganization that is clearly discontinuous from normality. Much as is the case with mental retardation, schizophrenia will probably ultimately be seen as a syndrome that is produced by multiple etiologies, potentially ranging from purely genetic to purely environmental and from neurodevelopmental to degenerative. If that working hypothesis is reasonable, then the most logical strategy for the study of schizophrenia is to use a relatively broad definition of the syndrome. Careful and comprehensive evaluation of clinical symptoms must be joined to careful and comprehensive evaluation of neuroimaging measures that may be useful in explaining pathophysiology and etiology. The same line of reasoning applies to the study of bipolar disorder, major depression, and their related spectrum conditions. Further, even the boundary between schizophrenia spectrum and affective spectrum disorders is unclear, as is the boundary between psychotic and nonpsychotic disorders within these spectra. Within this context, it is better to admit our ignorance and make such topics an object of investigation.

The Boundaries of the Concept and the Heterogeneity of the Schizophrenia Spectrum

The study of the relationship of "core schizophrenia" to affective disorders and other schizophrenia spectrum disorders illustrates some of the ways that breadth and heterogeneity issues can affect decisions about experimental design and the variable conclusions that can be drawn from studies depending on the nature of the "schizophrenic" sample studied. While the recent emphasis in current American clinical and research work has been to use a relatively narrow definition of schizophrenia, this approach may lead to the loss of true cases with only mild or relatively brief symptoms, thereby complicating identification of patterns of transmission and genetic markers, making it difficult to identify samples of first-episode patients, and essentially limiting the study of schizophrenia to patients with established chronicity. I believe that this approach should be carefully re-evaluated within the context of what we have recently learned from neuroimaging research (as will be discussed in the second half of this chapter). A few examples of boundary and heterogeneity issues illustrate the problems involved, based on evidence from epidemiologic, genetic, and treatment research.

Schizoaffective Disorder

Some patients with schizophrenia also have affective symptoms such as dysphoric mood, and the negative symptoms of schizophrenia are sometimes difficult to distinguish from depressive symptoms. Early studies indicated that intact affect and the presence of depressed mood were good prognostic signs in schizophrenia (Vaillant 1964; Stephens et al. 1966). It is highly likely that many of the patients in these studies would now be diagnosed as having schizoaffective disorder or affective disorder with mood-incongruent or even mood-congruent psychotic features. When lithium became widely available as a treatment for mania, clinicians began to use it for schizoaffective disorder as well, and found that some patients with mixed affective and schizophrenic symptoms responded well to lithium. Thereafter, the whole concept of schizoaffective disorder was re-evaluated; this research has now been summarized in a large number of review articles (Clayton 1984; Coryell et al. 1985). Many of these studies are flawed, however, by the fact that they pool many different definitions of schizoaffective disorder – some emphasizing acuteness of onset, others response to treatment, and others mixed phenomenology. Brockington et al. (1979) and others have summarized the many different criteria that have been applied to define schizoaffective disorder.

When this body of studies is reviewed critically, the most judicious conclusion is probably that the current definitions of schizoaffective disorder identify a mixture of atypical patients with schizophrenia and with affective disorder. For example, the studies of outcome usually indicate that schizoaffectives do less well than patients with pure affective disorder, but better than patients with schizophrenia. Studies of treatment indicate that schizoaffectives are less likely to respond to lithium alone, but do respond relatively well to combination therapy and show a better overall response than patients with pure schizophrenia. Family studies indicate that patients with schizoaffective disorder have higher rates of schizophrenia in their families than do patients with pure schizophrenia. All of these findings suggest that the term "schizoaffective disorder" refers to a heterogeneous group of patients, but this group must include at least some who belong within the schizophrenia spectrum.

There is no definition that is universally agreed upon for schizoaffective disorder. The DSM-III-R or DSM-IV definition stresses the persistence of "schizophrenia" symptoms after affective symptoms have remitted, while the ICD-10 definition stresses the simultaneous occurrence of both affective and schizophrenic symptoms. The former uses a longitudinal approach and the latter a cross-sectional approach; many of the patients identified as having schizoaffective disorder by ICD-10 would fall in the "mood-incongruent" category of mood disorders in DSM-III-R or DSM-IV. As the above review indicates, traditional validating measures – outcome, response to treatment, and familial aggregation – do not really solve the problem.

Given this state of uncertainty, the boundary between schizophrenia and affective disorder must therefore remain flexible, depending on the type of question being pursued. A narrow definition of schizophrenia allows for minimal contamination of schizophrenic cohorts by patients with affective disorder, although it may produce contamination of samples of patients with affective disorder by including some who, in fact, have schizophrenia. A narrow definition of schizophrenia may produce some problems for research in schizophrenia as well. In particular, genetic and family studies may benefit from a broader definition, since studies of diagnostic concordance among twins or of transmission within families can be flawed and lead to falsely elevated estimates of discordance or a failure to identify existing markers or patterns of transmission. In recognition of this problem, twin and linkage studies conducted to date have usually included schizoaffective patients as "cases," and some have even included patients with psychotic affective disorder as well.

Schizotypal Disorder

The issue of schizoaffective disorder addresses the boundary of schizophrenia with respect to affective symptoms, while the issue of schizotypal disorder addresses this boundary with respect to severity. In the preneuroleptic era, mild nonpsychotic forms of schizophrenia were gradually added to the concept under a variety of different names: latent, pseudoneurotic, and simple schizophrenia. As the risk of tardive dyskinesia became apparent, and as the Soviet abuse of disidents under the rubric of "sluggish schizohrenia" began to arouse international concern, pressure mounted to narrow the concept of schizophrenia in order to prevent excessive or inappropriate use of neuroleptics or the use of diagnosis as a means for political control. The current narrowing of the concept of schizophrenia in terms of severity arose for clinical, pharmacological, and political reasons, not because "mild" (i.e., nonpsychotic, nonpositive, or negative) schizophrenia was found to be discontinuous from severe schizophrenia.

Recent genetic and family studies suggest that the boundaries may currently be set too narrowly. From an etiological or pathophysiologic aspect, schizophrenia may exist as a spectrum of disorders including milder "personality" disorders (Kety et al. 1971). There are a number of studies which examine schizoid or schizotypal traits in twins, adopted offspring, or first-degree relatives of schizophrenic probands. In a re-analysis of the Danish adoption study of schizophrenia, Kendler and colleagues (1984) found that biologic relatives of adoptees with schizophrenia were at higher risk for schizotypal disorder than were relatives of adoptees without schizophrenic psychosis (14.3% versus 0%). An increased incidence of schizotypal disorder in the first-degree relatives of schizophrenics has also been reported (Kendler et al. 1984).

Schizotypal features are also linked phenomenologically to schizophrenia. Here, again, definitional issues are unclear and controversial, suggesting the need for closer examination and critical evaluation. For example, five of the eight criteria for schizotypal personality in DSM-III-R are nearly identical to the criteria for prodromal schizophrenia, suggesting a continuity between two disorders that the nosology implies to be diagnostically discrete. The importance of negative or defect symptoms (which are closely allied phenomenologically with schizotypal features and in their pure form define simple schizophrenia) also suggests a potential continuity between mild or nonpsychotic forms of schizophrenia and "core" chronic schizophrenia. Research on negative symptoms consistently indicates that these symptoms have important potential validity in identifying a subgroup of patients with cognitive impairment, poor premorbid features, or structural brain abnormalities. Thus, schizotypal traits and predominantly negative schizophrenia may share possible etiological links, as suggested by studies of ventricular size using neuroimaging (Schulsinger et al. 1984).

Identifying and Describing Characteristic Features

Given the state of the confusion described above, it is probably best to conceptualize the schizophrenia spectrum in terms of a relatively broad range of diagnoses. There are seven diagnoses that may be considered to belong within this spectrum. They include "*core*" *schizophrenia*, as defined by the narrow definition of DSM-IV, as well as the briefer form of the same syndrome, *schizophreniform disorder*, which differs only in terms of duration. The schizophrenia spectrum should probably also include two nonpsychotic illnesses that have traditionally been associated with the broad concept of schizophenia: *schizotypal disorder* and *simple schizophrenia* (which has reemerged in the DSM-IV appendix as simple disintegrative disorder and is present in ICD-10 under its original historic name). In addition, the schizophrenia spectrum almost certainly includes *schizoaffective disorder*, which has both depressive and manic forms. It should probably also include at least the psychotic forms of *mania* and *depression*, since mood-incongruent psychotic mania and depression are both essentially identical to older concepts of schizophrenia. These two severe affective syndromes share many phenomenologic features in common with core schizophenia, differing primarily in their episodic nature. It is not as yet clear whether the mechanisms that produce the psychotic symptoms in mania and depression are similar or identical to those producing the psychotic symptoms of schizophrenia, or whether the symptoms are a consequence of a different mechanism. Although the waxing and waning of psychotic symptoms in affective disorder suggests a mechanism operating at a molecular or cellular level, considerable evidence has been amassed from neuroimaging (discussed below) to suggest that these illnesses also have structural anatomic lesions.

In the late 1970s and early 1980s, Crow (1980) and Andreasen (1982) both suggested the possible utility of conceptualizing the characteristic features of schizophenia and its related spectrum conditions in terms of positive and negative symptoms. Crow proposed a model that involved two categories or syndromes, positive and negative, while Andreasen proposed adding a third mixed group. While this approach was partially supported by exploration of a variety of external validators (e.g., premorbid functioning, ventricular enlargement seen using CT, response to treatment), it is clearly at best a heuristic oversimplification. More recent work using factor-analytic techniques has suggested that the positive symptoms should be further subdivided into a psychotic and a disorganized group, producing three general dimensions of psychopathology: psychotic, disorganized, and negative (Bilder et al. 1985; Liddle 1987; Arndt et al. 1991; Miller et al. 1993). Although this dimensional approach

Table 1. Dimensions: defining symptoms

Psychotic
Delusions
Hallucinations
Disorganized
Disorganized speech
Disorganized behavior
Inappropriate affect
?Attention
Negative
Alogia
Affective blunting
Anhedonia
Avolition
?Attention

Table 2. Dimensional-diagnostic cross-walk

	Core	S-form	S-type	Simple	SAD	Mania	Dep
Psychotic	X	X			X	X	X
Disorganized	X	X			X	X	
Negative	X	?	X	X	?		X

to conceptualizing the symptoms of the major psychoses has been supported by numerous factor-analytic studies, as well as by some work using PET done by Liddle and others, it clearly needs further exploration. One recent factor-analytic study referred to positive (psychotic, disorganized) and negative dimensions of the schizophrenia spectrum. Based on such factor-analytic studies, a consensus is emerging as to which of the various symptoms of schizophrenia should be assigned to these three dimensions. One widely accepted solution appears in Table 1.

Very little attention has been given to date to the cross-walk between these dimensions and more conventional categorical diagnostic concepts (Table 2). Clearly, the dimensions cut across diagnostic categories. Although these dimensions have not been evaluated in other disorders apart from core schizophenia, one can hypothesize as to what would be found. For example, the psychotic dimension must, almost by definition, play a prominent role in the cross-sectional symptoms of schizophreniform disorder, schizoaffective disorder, and psychotic mania and depression. The disorganization dimension may occur in schizophreniform disorder, schizoaffective disorder, and mania as well, although it seems unlikely to be present in depression. Negative symptoms, on the other hand, are of questionable significance in schizophreniform disorder, if one conceptualizes it as a disorder that ends prior to a 6-month duration, although they could be a prominent part of schizophreniform disorder, if it is seen as

Table 3. The symptom spectrum

	Core	S-form	S-type	Simple	SAD	Mania	Dep
Delusions	X	X			X	X	X
Hallucinations	X	X			X	X	X
Disorganized speech	X	X			X	X	
Disorganized behavior	X	X		?	X	X	
Inappropriate affect	X	X			X		
Attention	X	X			X	X	X
Alogia	X	?	X	X	X		X
Affective blunting	X	?	X	X			X
Anhedonia	X	?	X	X			X
Avolition	X	?	X	X			X

merely a "holding pattern" before chronicity is sufficiently established in order to make a diagnosis of core schizophrenia. Negative symptoms are clearly prominent in both schizotypal disorder and simple schizophrenia, and they are also relatively prominent in depression, although with perhaps a different flavor. They may also be present in the depressive form of schizoaffective disorder.

Table 3 shows these relationships in a more fine-grained way, listing the various symptoms of schizophrenia in detail and showing the extent to which they are prominent in the various diagnoses within the schizophrenia spectrum. The pattern is illuminating, since it indicates that some components of the schzophrenia spectrum, such as schizoaffective disorder, mania, and depression, share prominent psychotic symptoms with core schizophrenia, while others share the symptoms at the negative end of the continuum. The patterning of symptoms within these disorders may give us some clue as to the different underlying neural mechanisms that produce these syndromes. When coupled with the fact that some symptoms appear in both core schizophenia and mania (e.g., disorganized speech or delusions), but are very reversible and treatment responsive in mania and relatively chronic in core schizophrenia, still more information is added that provides grist to the mill for hypothesizing about underlying mechanisms.

Brain Function and the Schizophrenia Spectrum

As we attempt to conceptualize the commonalities and differences of the neural substrates that underlie these various conditions, we should remind ourselves that the workings of the brain are as complex and interrelated as are the components of the schizophrenia spectrum. There are multiple levels of

Table 4. Sample models or hypotheses

Dopamine excess ⇒ Psychotic syndromes
Disrupted attentional circuitry ⇒ Disorganization syndromes,? Psychotic syndromes
Diminished frontal function ⇒ Negative symptoms
Superior temporal gyrus dysfunction ⇒ Auditory hallucinations

brain function and dysfunction which we must consider, including molecules, membranes, cells, circuits, and systems. Rapid online processing occurs through molecules, which communicate across membranes. The molecules are very fluid, while the membranes are at least somewhat plastic and modifiable (i.e., calcium channels, receptor induction, reuptake mechanisms, etc.). Cellular organization is somewhat more permanent, although cells can obviously die and drop out, and their axons can also grow back after injury, or they may form new dendritic connections. Circuits are more hard wired, although they are certainly modulated through molecular and membrane effects. Systems, typically conceptualized as corresponding to particular anatomic regions (e.g., motor systems in the basal ganglia or motor strip, language systems in the temporal lobes), are probably the most hard wired of these various levels.

Most of the working models currently being explored with neuroimaging techniques and other approaches to the study of the neural substrates of psychopathology draw on hypotheses about circuits or systems. Circuit models focus most of their mental efforts on particular anatomic regions (e.g., frontal heteromodal association cortex, temporolimbie cortex, basal ganglia circuitry, and reticular-thalamic-cingulate circuitry), while systems models work from a cognitive psychology orientation and emphasize the evaluation of cognitive systems, such as language, attention, or memory, and motor and "executive" elements.

Table 4 illustrates how some of these concepts have been integrated to produce particular models or hypotheses that can be explored using neuroimaging techniques. The dopamine hypothesis, which has been in circulation for nearly two decades, remains one of the most robust working models to explain psychotic syndromes. Although the dopamine hypothesis is said to be a hypothesis of schizophrenia, it is more properly conceptualized as a hypothesis for psychosis. Neuroleptic drugs that block dopamine transmission are most successful in reversing delusions and hallucinations, and they do so equally well in core schizophrenia, mania, or depression. Thus, if one postulates a dopaminergic mechanism, one must see it as applying to the psychotic dimension, but working less well for the full array of core symptoms that characterize schizophrenia. This example illustrates, however, the utility of conceptualizing at broader levels than diagnosis and employing the dimensional concept. A second model, which has been much less intensively explored, postulates that attentional circuitry is disrupted in psychosis. Attentional abnormalities were originally invoked by Broadbent and others to explain "thought disorder." This model could apply equally well to explaining at least some psychotic syndromes, in that the hypothesis suggests that the basic disorder involves excessive arousal and inability to suppress irrelevant input. Again, this hypothesis must also move across the broad spectrum, explaining symptoms of mania and schizoaffective disorder as well as core schizophrenia.

Another theory widely discussed during the past decade has been the "hypofrontality hypothesis." Although this hypothesis is often invoked in order to explain negative symptoms, neuroimaging studies tend to find hypofrontality in both core schizophrenia and affective syndromes, including syndromes such as mania that lack prominent negative symptoms.

As the final example in Table 4 illustrates, models can become extremely focused in both cortical localization and symptom type. A model that has been explored in several studies in recent years has involved postulating a dysfunction in the superior temporal gyrus to explain auditory hallucinations (Barta et al. 1990); abnormalities in the temporal region (i.e., hippocampus) have also sometimes been postulated to explain "thought disorder." Although these symptoms occur in affective syndromes as well as in core schizophrenia, examination of this model has not as yet been pursued in the affective illnesses.

Perspectives from Neuroimaging

Since the first report of the application of modern neuroimaging techniques to the study of schizophrenia in 1976 (Johnstone et al. 1976), several hundred studies have been reported concerning structural abnormalities in core schizophrenia. The early CT work focused primarily on the ventricular system and provided evidence to suggest that some type of brain abnormality was occurring in schizophrenia, although the mechanistic nonspecificity of ventricular enlargement provides very little definitive information concerning pathophysiology or level of abnormality.

Most recent assessments with MR techniques have provided a more refined method to evaluate brain structure, and these studies have sharpened the focus of the analysis. In addition to confirming ventricular enlargement, these studies have reported decreases in overall cerebral volume, as well as decreased cranial volume in a few studies, suggesting a dysgenetic developmental process. Other studies have reported increases in basal ganglia volumes and suggested that these findings represent a development failure in the normal "pruning process" in this region. Decreased frontal, temporal, and superior temporal gyrus size has also been reported, and in some studies the frontal findings have been linked to negative symptoms and the superior temporal gyrus findings to hallucinations or formal thought disorder. Finally, twin studies have indicated that the ill twin in discordant monozygotic pairs tends to have more severe abnormalities of a variety of types, results which have been interpreted to suggest that both DNA-driven and external environmental processes may play a role in the development of schizophrenia. All the MR studies have had a variety of methodological problems, such as small sample sizes, use of early-generation scanners, or poorly validated measurement techniques. In addition, a variety of methodological issues are inherent in MR research that are as yet not fully resolved, such as the best way to control for body size or head size, the effects of subject or parental education, and relationships of brain measures to intelligence; these issues all affect the choice and interpretation of control samples. Consequently, none of the important or suggestive studies completed to date can be considered to be definitive. MR studies, while they have contributed a great deal to our knowledge of core schizophrenia, must still be regarded as in the "hypothesis-generation phase" concerning the neural substrates of symptoms, dimensions, or diagnosis. A summary of these findings has been provided in several recent reviews (Andreasen et al. 1993; Pearlson and Marsh 1993).

The functional imaging techniques appear to suggest a greater consensus concerning the types of abnormalities observed in core schizophrenia. In this instance, the situation is quite similar to the finding of ventricular enlargement in schizophrenia assessed with CT. The majority of studies using both SPECT (or related measures of cerebral blood flow with cortical probes) and PET have looked for and found "hypofrontality." Although a number of negative studies have been reported, the majority have been positive. Early work emphasized single studies of "the resting state," but more recently studies have used the more powerful strategy of comparing a baseline condition to an experimental condition that "challenges" the frontal lobes to respond to a carefully selected "frontal lobe task" (Weinberger et al. 1986). A recent study, focusing on first-episode, never-treated patients, has also suggested that these findings are not due to long-term effects of medication or chronicity and that they may be related to negative symptoms (Andreasen et al. 1992). One set of studies by Liddle et al. (1992) has explored regional abnormalities with PET in relation to dimensional approaches, but the majority of functional imaging studies have remained at the diagnostic level. Given that the functional imaging techniques, and especially the ^{15}O water method, offer an especially powerful strategy for disentangling components of cognitive systems and neural circuitry and examining relationships between these levels of brain function and psychopathology, an open frontier is still before us for the investigation of cerebral physiology and schizophrenia. Work in this area completed to date has also been summarized in several recent reviews (Buchsbaum 1993; Prohovnik 1993).

The bulk of the neuroimaging work completed to date has focused on the topic of core schizophrenia. If we widen the field of view and examine the entire schizophrenia spectrum, as defined above, the enlargement of the field blurs the focus considerably. Nearly all studies have worked at the diagnostic level. The majority have examined the conditions at the opposite end of the spectrum: affective disorders. There have been very few studies of other spectrum conditions, such as schizophreniform disorder, schizotypal disorder, or schizoaffective disorder.

Early studies of affective disorder used CT scanning. Although the data are less definitive than for core schizophrenia, at least half of the studies of both bipolar and depressed patients have found evidence of ventricular enlargement (Andreasen et al. 1990). These CT studies are particularly plagued by methodological problems, with some even failing to clarify whether affective samples are manic or depressed. A survey of MR studies indicates a similar pattern. Although the number of studies using MR to date are somewhat limited, the majority have reported evidence for structural abnormalities, including ventricular enlargement. Some have examined more specific brain regions; as in the case of core schizophrenia, abnormalities have been reported in temporolimbic regions, hippocampus, and basal ganglia (Pearlson and Marsh 1993; Swayze et al. 1990, 1992). The only type of structural abnormality which to date has shown a clear dissociation between affective disorder and core schizophrenia is focal signal hyperintensities in subcortical white matter areas, referred to colloquially as "unidentified bright objects" or "UBOs." This finding has been relatively consistently reported both in bipolar patients and in elderly depressed patients, but a significantly increased rate of UBOs has not been observed in schizophrenia. The patholophysiological significance of this finding is unclear, but it may point to a single brain mechanism for which these two conditions (i.e., core schizophrenia and affective disorder) differ. Focal signal hyperintensities are seen in demyelinating disorders and in cardiovascular disorders such as hypertension (leading to small areas of infarction in the brain). Although the findings in depression have been limited to older patients, samples of relatively young bipolar patients have been studied, and in these samples the effects of treatment or a history of hypertension do not appear to contribute to the finding.

Functional imaging studies of affective disorder show perhaps the most stiking similarity to studies of core schizophrenia. Buchsbaum (1993) recently reviewed this topic. He summarizes nine studies using PET completed between 1984 and 1990. All studies showed an overall decrease in glucose utilization in patients with affective illness, and seven of the nine also showed hypofrontality, which may be more prominent in bipolar than unipolar patients. These findings are also consistent with the lesion studies of Robinson et al., which suggest that disruptions of input to the frontal cortex produced by strokes may produce an affective syndrome, with differential laterality effects producing a manic or a depressive syndrome (Robinson et al. 1984).

Neuroimaging studies of the "middle" of the schizophrenia spectrum are notably skimpy. A few, largely using CT, have documented a pattern of ventricular enlargement in schizophreniform disorder or schizoaffective disorder that appears to be similar to core schizophrenia (Rieder et al. 1983; Weinberger 1984).

Turning to the portion of the "middle group" that lacks severe psychotic symptomatology, we find even less. There are two published studies of schizotypal disorder. The first report, by Schulsinger et al. (1984), examined a high-risk sample of children of schizophrenic mothers who were divided into three groups: those who developed schizophrenia ($n = 7$), "borderline schizophrenics" ($n = 11$), and those with no mental illness ($n = 13$). Compared with normal controls, the individuals who developed schizophrenia showed a significantly greater VBR measured with CT, while the borderline group had smaller ventricles than the controls and the normal offspring were in a middle position.

They interpreted this finding as consistent with a model that suggested core schizophrenia and schizotypal disorder to share the same genetic diathesis, but with an environmental insult being required to precipitate the actual development of schizophrenia. A second study by Raine et. al (1992) examined schizotypal disorder in a population of normal subjects using MR, neuropsychological tests of frontal lobe function, and various measures of schizotopy. They observed high negative correlations between the measures of schizotopy and MR measures of left and right frontal lobes and similar correlations between the measures of schizotopy and performance on frontal lobe neuropsychological tasks. Although it focuses on normals, this study is an example of taking a dimensional approach to examining brain–behavior relationships.

We recently completed a small study of schizotypal disorder using MR. This study was a subset of an ongoing series of MR studies being completed at the University of Iowa Mental Health Clinical Research Center. As the third in our series, this particular study is referred to as "MR3."

MR3 is one of the largest studies comparing a sample of patients suffering from core schizophrenia ($n = 102$) with healthy normal volunteers ($n = 87$) that has been completed to date. Notable findings in the overall study are an increase in ventricular size and a decrease in thalamic size in the schizophrenic patients, after gender, height, and age have been controlled for through an analysis of covariance. There is a trend toward a decrease in size of the hippocampus and superior temporal gyrus as well. These results are summarized in Table 5. In addition to the large sample of core schizophrenics, we have studied a small sample of patients with schizotypal disorder ($n = 10$). We therefore compared the schizotypals with the normals and schizophrenic patients using similar corrections for age, gender, and height. Statistical power was limited due to the small sample size. Nevertheless, some patterns of trends were noticed which may be of potential interest. The schizotypals were more similar to the schizophrenics than the normal controls on the two measures that most robustly distinguish the schizoprenics from normals in the overall sample: lateral ventricles and thalamus. These measures reflect midline abnormalities. The thalamus in particular is a crucial structure that may mediate "gating" and "attention"; an abnormality in this region could explain some commonalities in clinical presentation between schizotypal disorder and core schizophrenia, such as prominent negative symptoms. On the other hand, the schizotypal patients were more similar to the normals in size of third ventricle, hippocampus, and superior temporal gyrus. While the third ventricular finding is consistent with the explanation mentioned above concerning midline circuitry, the "normality" of the hippocampus and temporal cortical structures may explain the absence of prominent psychotic features in indivudals with schizotypal disorder.

We also pursued a strategy similar to that of Raine and examined correlations between the Chapman Perceptual Aberration and Physical Anhedonia Scales and the following brain regions: left and right frontal, left and right prefrontal, left and right cerebrum, cranium, cerebellum, left and right hippocampus, left and right lateral ventricle, and third ventricle. Very few of the correlates were statistically significant, and the findings of Raine concerning a relationship

Table 5. "MRI 3" results: schizophrenia (n = 102) vs. controls (n = 87)

Structure (ROI)	DX
Third ventricule	7.66***
Lateral ventricles	5.62**
Cranium	0.26
Cereberum	0.29
Cerebellum	1.60
Temporal lobes	1.15
Sup. temp. gyrus	3.48*
Temporal horns	1.20
Hippocampus	3.62*
Caudate	0.11
Thalamus	4.58**

*$p < .10$; **$p < .05$; ***$p < .01$

between schizotopy measures and frontal abnormalities were not confirmed. The most robust correlations observed were a negative relationship between left frontal size and the Perceptual Aberration Scale (−.2) and a positive correlation between physical anhedonia and right prefrontal regions (.15). Significant correlations were noted between cerebellar size and physical anhedonia ($r = -.32$, $p \leq .01$) and between perceptual aberration and left hippocampal size ($r = -2.6$, $p \leq .05$). The hippocampal results are consistent with models positing a relationship between temporolimbic structures and perceptual aberration. The cerebellar findings, on the other hand, are more difficult to explain; they may reflect a real, but rarely investigated, role for the cerebellum as a modulator of higher cognitive behavior. This finding is consistent with out own work showing a relationship between cerebellar size and general intelligence (Andreasen et al. 1993).

Summary and Conclusion

Studies using neuroimaging techniques have focused primarily on extreme ends of the schizophrenia spectrum. This strategy makes sense because of the costs and technological complexity involved in such techniques. When one is still exploring and attempting to find one's way on difficult terrain, it is usually wise to begin with the easiest and clearest path. Avoiding blurry diagnostic issues has made sense, therefore, up to a point. In spite of the efforts to obtain a clear answer, definitive findings have not as yet been forthcoming from the various neuroimaging studies conducted to date. The ambiguity and nonreplicability of findings up to this point probably do not reflect deficiencies in either the studies themselves or the techniques, but rather the inherent difficulty of the task.

Surprisingly, when the studies examining the extreme ends of the spectrum are compared with one another, the results are surprisingly similar. Patients with core schizophrenia and patients with severe affective syndromes have relatively similar findings, whether the imaging technique is CT, MR, or PET. The implications of these results are not clear, but they suggest that dimensional approaches to studying psychopathology may be more reflective of clinical reality and underlying neural mechanisms than are the categorical diagnostic approaches that have traditionally been used. A modest survey of the "middle" of the schizophrenia spectrum also supports this conclusion. Increasingly, researchers should investigate brain–behavior relationships not only in terms of diagnosis, but also in terms of dimensions and of specific symptoms.

References

Andreasen NC (1982) Negative symptoms in schizophrenia: definition and reliability. Arch Gen Psychiatry 39:784–788

Andreasen NC, Swayze V, Flaum M, Yates WR, Arndt S, McChesney C (1990) Ventricular enlargement in schizophrenia evaluated with CT scanning: effects of gender, age, and stage of illness. Arch Gen Psychiatry 47:1054–1059

Andreasen NC, Rezai K, Alliger R, Swayze VW, Flaum M, Kirchner P, Cohen G, O'Leary DS (1992) Hypofrontality in neuroleptic-naive and chronic schizophrenic patients: assessment with xenon-133 single-photon emission computed tomography and the Tower of London. Arch Gen Psychiatry 49:943–958

Andreasen NC, O'Leary DS, Arndt S (1993) Neuroimaging and clinical neuroscience: basic issues and principles. In: Andreasen NC (ed) Review of psychiatry. American Psychiatric Press, Washington, DC, pp 315–346

Arndt S, Alliger RJ, Andreasen NC (1991) The distinction of positive and negative symptoms: the failure of a two-dimensional model. Br J Psychiatry 158:317–322

Barta PE, Pearlson GD, Powers RE, Richards SS, Tune LE (1990) Auditory hallucinations and smaller superior temporal gyral volume in schizophrenia. Am J Psychiatry 147:1457–1462

Bilder RM, Mukherjee S, Rieder RO, Pandurangi AK (1985) Symptomatic and neuropsychological components of defect states. Schizophr Bull 11:409–491

Bleuler E (1911) Translated by J. Zinkin (1950) Dementia praecox of the group of schizophrenias. New York International Universities Press, New York

Brockington IF, Wainwright S, Kendell RE (1979) Schizoaffective psychoses: definition and incidence. Psychol Med 9:91–99

Buchsbaum MS (1993) Positron-emission tomography and brain activity in psychiatry. In: Andreasen NC (ed) Review of psychiatry. American Psychiatric Press, Washington, DC, pp 461–486

Clayton P (1984) Schizoaffective disorders. J Nerv Ment Disord 170:646–650

Cooper JE, Kendell RE, Gurland BJ, Sharp L, Copeland JRM, Simon R (1972) Psychiatric diagnosis in New York and London (Maudsley Monograph no. 20). Oxford University Press, London

Coryell W, Endicott J, Keller M, Andreasen NC (1985) Phenomenology and family history in DSM-III psychotic depression. J Affective Disord 9:13–18

Crow TJ (1980) Positive and negative schizophrenic symptoms and the role of dopamine. Br J Psychiatry 137:383–386

Johnstone EC, Frith CD, Crow TJ, Husband J, Kreel L (1976) Cerebral ventricular size and cognitive impairment in chronic schizophrenia. Lancet 2:924–926

Kendler KS, Materson C, Ungaro R, Davis K (1984) A family history study of schizophrenia-related personality disorders. Am J Psychiatry 141:424–427
Kendler KS, Tsuang MT (1984) Outcome of schizophrenic subtypes defined by four diagnostic systems. Arch Gen Psychiatry 41:149–154
Kety SS, Rosenthal D, Wender PH, Schulsinger F (1971) Mental illness in the biological and adoptive families of adopted schizophrenics. Am J Psychiatry 128:302–306
Kraepelin E, Barclay RM, Robertson GM (1919) Dementia praecox and paraphrenia. E&S Livingstone, Edinburgh
Liddle PF (1987) The symptoms of chronic schizophrenia. A re-examination of the positive-negative dichotomy. Br J Psychiatry 151:145–151
Liddle PF, Friston KJ, Frith CD, Frackowiak RSJ (1992) Cerebral blood flow and mental processes in schizophrenia. J R Soc Med 85:224–226
Mellor CS (1970) First-rank symptoms of schizophrenia. B J Psychiatry 117:15–23
Miller DD, Arndt S, Andreasen NC (1993) Alogia, attentional impairment, and inappropriate affect: their status in the dimensions of schizophrenia. Compr Psychiatry 34:221–226
Pearlson GD, Marsh L (1993) Magnetic resonance imaging in psychiatry. In: Andreasen NC (ed) Review of psychiatry. American Psychiatric Press, Washington, DC, pp 347–382
Prohovnik I (1993) SPECT imaging of cerebral physiology. In: Andreasen NC (ed) Review of psychiatry. American Psychiatric Press, Washington, DC, pp 421–459
Raine A, et al (1992) Pre-frontal structural and functional deficits associated with individual differences in schizotypal personality. Schizophr Res 7:237–247
Rieder RO, Mann LS, Weinberger DR, Van Kammen DP, Post RM (1983) Computed tomographic scans in patients with schizophrenia, schizoaffective, and bipolar affective disorder. Arch Gen Psychiatry 40:735–739
Robinson RG, Kubos KL, Starr LB, Rao K, Price TR (1984) Mood disorders in stroke patients: importance of location of lesion. Brain 107:81–93
Schulsinger F, Parnas J, Petersen ET, Schulsinger H, Teasdale TW, Mednick SA, Mthetaller L, Silverton L (1984) Cerebral ventricular size in the offspring of schizophrenic mothers. A preliminary study. Arch Gen Psychiatry 41:602–606
Stephens JH, Astrup C, Mangrum JC (1966) Prognostic factors in recovered and deteriorated schizophrenics. Am J Psychiatry 122:1116–1121
Swayze VW II, Andreasen NC, Ehrhardt JC, Alliger R, Cohen G (1990) Developmental abnormalities in schizophrenia. Arch Neurol 47:1054–1059
Swayze VW II, Andreasen NC, Alliger R, Yuh WTC, Ehrhardt JC (1992) Subcortical and temporal structures in affective disorder and schizophrenia: a magnetic resonance imaging study. Biol Psychiatry 31:221–240
Saillant GE (1964) Prospective prediction of schizophrenic remission. Arch Gen Psychiatry 11:509–518
Weinberger DR (1984) Computed tomography (CT) findings in schizophrenia: speculation on the meaning of it all. J Psychiatr Res 18:477–490
Weinberger DR, Berman KF, Zec RF (1986) Physiological dysfunction of dorsolateral prefrontal cortex in schizophrenia. I. Regional cerebral blood flow (rCBF) evidence. Arch Gen Psychiatry 43:114–124
Wing JK, Cooper JE, Sartorius N (1974) The measurement and classification of psychiatric symptoms. Cambridge University Press, Cambridge

Biochemical Heterogeneity in Schizophrenia: Implications and Research Strategies of the State Dependency Model

D. P. van Kammen

Introduction

Clinical heterogeneity in schizophrenia has limited interest unless we can validate it in terms of etiology, prognosis, or treatment response (Garver et al. 1988; Brown et al. 1990). Similarly, biochemical heterogeneity is not very interesting if it does not lead to etiological heterogeneity or clinical application. Traditional research strategies have been confounded by the fact that the study of schizophrenic patients is very difficult. The very nature of the disorder interferes with the patients' willingness and ability to participate in research (and in clinical care). Therefore, most studies make use of cross-sectional study designs, based on the notion that the study variable is stable over time. The difficulties in replicating such findings are usually explained by invoking the heterogeneity in etiology hypothesis, which explains much but clarifies nothing.

Most biochemical studies of schizophrenia have focused on potential differences between schizophrenic patients and control subjects, in the hope of finding the etiology of schizophrenia. Given that schizophrenia is such a frequent disorder with heterogeneous presentation, course, and drug response, heterogeneity in etiology is very likely (Slater 1947). Because the variability in biochemical measures has been the hallmark of schizophrenia research, the existence of biochemical subtypes with different etiologies seems a reasonable explanation. The early observation that all antipsychotic drugs share the ability to acutely block dopamine transmission (van Rossum 1966; Carlsson and Lindqvist 1963) created great hopes for finding such subgroups, which culminated in the heuristic division of schizophrenia into types I and II (i.e., dopaminergic and nondopaminergic) (Crow et al. 1980).

So far, subgroups based upon monoamine measures have not been found. There may be several reasons for this. In addition to potential phenocopies that we have not identified as yet, heterogeneity in clinical presentation of schizophrenia may reflect pleiotropism (McKusick 1969), which often creates diagnostic confusion. Examples of pleiotropism are a single abnormal or missing gene with multiple effects, e.g., phenylketonuria gene (Penrose 1952) and nongenetic medical conditions with a specific pathogenic process and multiple

---

Professor of Psychiatry, Department of Psychiatry, University of Pittsburgh School of Medicine, Chief of Staff, Highland Drive VAMC, Pittsburgh, PA 15206, USA

clinical manifestations, such as syphilis or systemic lupus erythematosus. What we diagnose as schizophrenia may be only one of the varied presentations of the disorder. Family studies of schizophrenia are beginning to sort out what is being transmitted (Parnas et al. 1993; Holzman et al. 1988) and what is common among the schizophrenia spectrum disorders (Siever et al. 1990; Condray and Steinhauer 1992). From a research point of view, pleiotropism of a single etiology is more attractive than a multifactorial model. However, our failure to come up with meaningful subgroups suggests that perhaps there are no distinct subgroups: Is schizophrenia a disorder caused by a multifactorial polygenic process, and do the measures that we are studying reflect not etiology but pathophysiology?

In this paper I will review the evidence, or rather the lack of evidence, of biochemical subgroups in schizophrenia, focusing on the monoamine systems, and the potential reasons for our failure to come up with such subgroups. I will discuss the concepts behind different research strategies. I will review drug response and relapse prediction studies, with an emphasis on our interim relapse prediction models. Such studies are based on the notion that variability in behavioral and biochemical measures during haloperidol treatment may be state-dependent relapse prodromes, which identify relapse risk after haloperidol discontinuation.

Are There Biochemical Subgroups in Schizophrenia?

The great expectation of the 1970s that a dopamine disturbance would emerge as the major etiology of schizophrenia has not materialized. For the dopamine system the positive studies range from elevated plasma HVA (Davidson et al. 1987) and decreased urinary HVA (Karoum et al. 1987) to no difference in CSF HVA (van Kammen et al. 1986a; Widerlöv 1988) and decreases in CSF HVA and increases in plasma HVA after neuroleptic withdrawal (Pickar et al. 1990; van Kammen et al., unpublished data, Nov. 1989). However, none of these findings have led to consistent identification of subgroups. Even the proposed dichotomy of dopaminergic versus nondopaminergic groups of schizophrenia (Crow et al. 1980) has not been confirmed (MacKay 1980). The only measure that may have a potential bimodal distribution is D_2 receptor binding in autopsy brains (Seeman et al. 1984). The PET scan studies of D_2 receptor binding in vivo in never-treated schizophrenic patients did not answer this question for certain (Farde et al. 1990; Wong et al. 1986; Crawley et al. 1986). No evidence has emerged, either, that the D_2 receptor gene is altered in schizophrenia (Moises et al. 1991).

Different etiologic hypotheses emerged, going back to Bleuler's concept of the group of schizophrenias, with the hope that different subtypes of schizophrenia could be identified. Besides the dopamine hypothesis of schizophrenia, specific biochemical hypotheses included serotonin (Gillin et al. 1976), norepinephrine (Stein and Wise 1971), gamma amino-butyric acid (GABA) (Roberts 1973; van Kammen 1977), endorphins (Terenius et al. 1976;

Verhoeven et al. 1979), the immune system (Torrey and Peterson 1976; Rapaport et al. 1989), and membrane phospholipids (Henn and Henn 1982; Yao and van Kammen 1994). Most of these hypotheses may still have some viability, but not even one single subgroup has been identified.

There may be reasons other than pleiotropism or the existence of phenocopies for this failure to identify biochemical subgroups in schizophrenia. First of all, as Tsuang et al. (1990) and Mortimer (1992) pointed out in their seminal papers, clinical researchers have not established whether any of the studied phenomena in schizophrenia are primarily or secondarily related to its etiology. An example on the clinical side are negative symptoms. Unless we define the deficit syndrome under rigid circumstances (Carpenter et al. 1988), we will run into the same problem as with biochemical variability. Negative symptoms are not as stable as is believed (Johnstone et al. 1986); they may be drug responsive, drug nonresponsive, or drug induced (van Kammen 1991) and secondary to primary changes in the disorder. Negative symptoms are best evaluated in unmedicated patients when psychotic symptoms are in remission (Kelley et al. 1992). Unless observed in a longitudinal design, such an approach could lead to false subdivisions: remitted or stable versus exacerbated instead of negative versus positive symptoms.

Another possibility is that the abnormality is hidden behind the variance we observe in our biochemical measures as opposed to significant mean differences from control groups. The lack of replication in biochemistry may be related to the variability over time as observed in symptomatology.

What Are the "Other" Causes of Biochemical Heterogeneity?

In order to reduce the variance through methodological differences and to increase the likelihood of replication, the patient sample needs to be well described: age, gender, diagnostic system, duration of illness, age of onset, medication status, dose and duration of treatment, premorbid functioning, positive and negative symptoms, description of scales and inter-rater reliability and relapse status. Other methodological issues include time of day and consistency of time of sample collection, diurnal and seasonal variations, time of drug-free clinical state, diet, activity level, duration and conditions of sample storage, assay descriptions and inter- and intra-assay reliability, and statistical methodology. The reader is referred to van Kammen and Sternberg (1980) and Post et al. (1983) for further discussions of these issues in CSF studies. Most of these sources of variance can and should be controlled for.

Is Treatment Nonresponse an Etiological Subtype?

Others, such as Brown and Herz (1987), have made a case that heterogeneity in drug response suggests heterogeneity in biochemical etiology, still based upon the assumption that the role of dopamine in antipsychotic drug response reflects biochemical etiology. Although it is a testable hypothesis, most of the data do not support it. Wolkin et al. (1989) showed that D_2 receptor binding with PET scanning is the same for responders and nonresponders. Conceivably, the lack of response results from a disturbance downstream of the receptor. There is also no evidence that drug response is stable over time (van der Velde 1976), although poor responders to neuroleptics are less likely to become good responders with time. Drug response does not run consistently in families with multiple schizophrenic probands (Delisi and Dauphinais 1989), making it unlikely that drug response is under genetic control, i.e., reflects etiological subgroups. The clozapine study of neuroleptic nonresponders has shown how hard it is to identify real nonresponders (Kane et al. 1988). A good drug response to clozapine by some poor responders to neuroleptics, which may not be repeated after clozapine treatment has been interrupted, questions the drug-response subtype even further. There is some evidence now that delay in treating the first episode may lead to a persistent poor treatment response (Loebel et al. 1992). Therefore, if drug response is a trait marker, it is in the direction of failure to respond, in full, early in the illness.

State-Dependent Markers of Drug Response

Drug response as a potential etiological marker needs to be separated from illness modifiers and from changes in state-dependent systems that have changed irreversibly. Garver et al. (1988) approached this issue in a different way; they examined rapid responders versus slow responders. However, the time to response may be controlled by processes unrelated to the etiology of schizophrenia. Similar to severity of illness, drug response seems to be on a continuum, precluding it from being a likely candidate for an etiological subgroup. The important issue here is whether these phenomena can be reliably and consistently observed and how we formulate that drug response relates to the etiology of psychosis or schizophrenia.

Several groups have reported that higher plasma HVA or MHPG leads to better antipsychotic drug response (Davila et al. 1988; Chang et al. 1990; Mazure et al. 1991), raising the hope for subgrouping. The assumption behind this is that a measure such as plasma HVA is a relatively stable characteristic, i.e., that a cross-sectional design will allow us to answer the question about a significant difference between subgroups of schizophrenic patients. Because plasma HVA and MHPG seem to be in part state dependent (Maas et al. 1993a,b), and higher psychosis levels tend to be associated with a better drug response (Angrist et al. 1981), drug response may also be state dependent.

What Are the Conceptual and Methodological Implications of Biochemical Heterogeneity?

There are different ways of looking at heterogeneity. The variance in our measures may be due to different etiologies or different expression thresholds for the illness (trait, diagnostic, or vulnerability markers), may have no relevance to schizophrenia, or may be related to intrinsic changes related to the disorder that are, however, secondary or tertiary level in origin, i.e., state-dependent changes (episode markers and epiphenomena). Many biological indices have been proposed as vulnerability factors. Vulnerability factors can be divided into stable (diagnostic markers and illness modifiers) and unstable (episode markers and mixed trait/state markers) types (Nuechterlein et al. 1992; Nuechterlein and Dawson 1984; van Kammen et al. 1982a; Zubin and Steinhauer 1981). Epiphenomena are also of interest, but traditionally less so; change in plasma HVA may be such an epiphenomenon because it trails the drug-induced behavior change. There is also variance in factors that are normally distributed and that are not etiologically related to schizophrenia but affect the form and outcome of the illness, the so-called illness modifiers. One example is lower dopamine β-hydroxylase (DBH) (Sternberg et al. 1983; van Kammen et al. 1994), which is associated with a more episodic illness, better premorbid functioning, and better drug response in schizophrenia. A bimodal distribution in CSF or plasma DBH is not present, nor is there bimodality in premorbid functioning. Platelet monoamine oxidase (MAO) activity has recently lost most of its attraction for study but could resurface as an illness modifier, be under state-dependent control, or have other pathophysiological implications, rather than as a potential etiological marker (Reveley et al. 1983). A longitudinal approach through remission and relapse, with and without medication, is required to determine the state of these markers (van Kammen et al. 1990; van Kammen and Kelley 1991; van Kammen 1991).

Although more expensive and more difficult to execute, longitudinal designs are needed to study drug effects. For example, in cross-sectional studies of medication effects, the nonmedicated patients may have a symptom intensity similar to that of the medicated patients (Gattaz et al. 1983). Minimally symptomatic drug-free patients may be invited to participate, because of the discomfort of many staff members with unmedicated patients and the difficulty of engaging very symptomatic unmedicated relapsed patients in study protocols, while medicated patients may be referred for study because of less than optimal drug response. Therefore, unjustified conclusions about drug effects may be drawn from cross-sectional studies. Furthermore, paired tests in longitudinal designs have greater statistical power, allowing for smaller groups of patients to suffice.

What Are the Basic Assumptions Dictating Our Study Paradigms?

The way we formulate our etiological models determines how we design our studies. The basic assumption behind our schizophrenia research is that schizophrenia is a chronic but episodic illness. The waxing and waning of symptoms may reflect a waxing and waning of biochemical processes. The neurochemical variations and problems of replication go beyond assay variability and clinical methodology used (van Kammen and Sternberg 1980; Post et al. 1983). Similarly to schizophrenic episodes being preceded by changes in behavior that can be detected prior to relapse, biochemical prodromes may be present. Therefore, biochemical variability suggests state-dependent episode markers but does not necessarily indicate different etiologies (although there still could be different etiologies). We choose the drug discontinuation paradigm in order to identify who is at high or low risk of relapse, because drug withdrawal is associated with an increased risk of relapse. The state-dependency model implies that relapse and drug response prediction may be possible, and that longitudinal designs are to be preferred over a cross-sectional design.

State-Dependent Markers and Catecholamine Activity

Examples of state-dependent or episode markers are CSF NE (van Kammen et al. 1989), MHPG (van Kammen et al. 1989; Pickar et al. 1990) and HVA (Pickar et al. 1990; Maas et al. 1993a,b; Bowers et al. 1986; Post et al. 1975), plasma HVA (Bowers et al. 1986; Davis et al. 1985; Davidson and Davis 1988; Pickar et al. 1990) and MHPG (Pickar et al. 1990; Ko et al. 1988; Bowers et al. 1986), and, potentially, urinary variables as well (Ko et al. 1988; Maas et al. 1993a,b). Plasma HVA seems to be higher, while urinary HVA is decreased (Karoum et al. 1987; Maas et al. 1993a,b). Interestingly, plasma HVA changes with neuroleptic treatment and after drug withdrawal seem to follow changes in behavior, while there is a significant relationship with psychosis ratings in the drug-free state (Pickar et al. 1990; Davis et al. 1991). It is conceivable that plasma HVA does not reflect brain dopamine activity, but centrally induced changes in the peripheral autonomic nervous system (Lambert et al. 1991). Exploratory studies of psychophysiology (e.g., Öhlund et al. 1992) and plasma HVA seem indicated. State dependency in D_2 or α_2 receptor activity has not been established.

Risk of Relapse

Risk of relapse is affected by (a) medication status, (b) dose of antipsychotic drug, (c) increased stress or life events (Ventura et al. 1992), and (d) changes in the intrinsic biochemical milieu (for review see van Kammen 1991). Because of the uncertainty about when relapse will occur, clinicians do not keep schizophrenic patients drug free. The present-day recommendation for relapse treatment and prevention suggests continued neuroleptic treatment for up to 1 year following the first episode, for up to 5 years after the second episode, and indefinitely after the third episode. A majority of patients, however, do not take medications as prescribed at any given time. Others still may relapse on neuroleptic treatment (Ventura et al. 1992). If relapse on medication is the result of increased stressors/life events and relapse without medication may occur with minimal or daily stresses, this indicates an increased stress sensitivity preceding the actual psychotic exacerbation (Nuechterlein et al. 1992; van Kammen and van Kammen 1984; Albus et al. 1982; Breier et al. 1989).

In spite of attempts to predict relapse in schizophrenia by using behavioral prodromes (Subotnik and Nuechterlein 1988; Dencker et al. 1986; Marder et al. 1979; Herz et al. 1982,1991; Heinrichs and Carpenter 1985), clinicians are not very successful in predicting who will remain in a stable clinical state without antipsychotic drugs (Johnson 1976; Johnson et al. 1983; Buchanan et al. 1992). On the other hand, not all potential "prodromal behaviors" are consistently followed by relapse within a reasonable time (Subotnik and Nuechterlein 1988), while some patients relapse precipitously (Herz et al. 1991). Behavioral prodromes may signal a potential relapse but do not have the required specificity or sensitivity to predict how long it will take before the psychotic exacerbation will occur. Actually, over 50% of patients relapse without obvious behavioral prodromes (Herz et al. 1991; Hirsch et al. 1990). This is clinically important, because many schizophrenic patients are not adhering to the prescribed medication regimens.

Most clinicians will respond to fluctuations in symptoms with pharmacological intervention, because of the belief that relapse is imminent (Dencker et al. 1986). If such interventions do not take place in a timely fashion, however, a full-blown psychotic relapse may disrupt the patients' functioning for a prolonged period of time (Herz et al. 1991). While our ability to improve our patients' lot by decreasing the stresses of daily life are limited, the effectiveness of our interventions is lessened by our lack of knowledge of the relapse process. Developing reliable predictors of clinical stability has obvious practical clinical implications. Studying relapse prediction will at the same time increase our understanding of the behavioral and biochemical processes involved in relapse.

Relapse Prediction: The Early Studies

Behavioral prodromes have been recognized for a long time (Buchanan et al. 1992; Nuechterlein et al. 1992; Hirsch et al. 1990; Herz et al. 1982; Docherty et al. 1978). Sleep has also been reported to decrease prior to relapse (van Kammen and Kelley 1991). The first reports of relapse prediction with biochemical measures focussed on dopamine activity; e.g., lower plasma prolactin was associated with early relapse following drug withdrawal (Brown and Laughren 1981), although recent studies did not seem to confirm the earlier fingings (Kirkpatrick et al. 1992; Green et al. 1990), and growth hormone (GH) but not prolactin response to apomorphine was greater in psychotic than in remitted patients (Müller-Spahn et al. 1985). Cleghorn and associates (1983) observed that the GH response to apomorphine was lower in drug-free schizophrenic patients than in normals but increased prior to relapse.

Variable responses to amphetamine or methylphenidate, as noted by Janowsky et al. (1973) and Janowsky and Davis (1976), could be explained as reflecting different types of schizophrenia. Repeated amphetamine infusions in the same patient indicated that the response was not always identical (van Kammen et al. 1982a,b,c). The psychotogenic effect of dopamine agonists indicated not only a dopamine hyper-responsiveness in schizophrenia, but that there also were different modifying factors. We noticed that patients on pimozide who showed a worsening were the same patients who subsequently relapsed after pimozide withdrawal, and we concluded that the relapse process was a dynamic process (van Kammen et al. 1982a). Two patients who showed an improvement improved also temporarily following pimozide withdrawal (van Kammen et al. 1980, unpublished data). Angrist et al. (1985, 1981) observed a similar association between worsening with amphetamine and following drug withdrawal.

The next series of dopamine agonist studies, i.e., of methylphenidate by the Lieberman group (1984, 1993), indicated that the psychotogenic response to methylphenidate also related to a more rapid relapse rate following drug withdrawal up to 1 year. Similarly, Davidson et al. (1987) gave L-dopa to stable schizophrenic patients for a week, and their increase in psychosis was also associated with subsequent relapse. The stimulant challenge test data (for review see Angrist and van Kammen 1984) indicated that whether patients worsen in response to a pharmacological challenge test or drug withdrawal may be determined by biochemical instability (Lieberman et al. 1984; van Kammen et al. 1982a, 1990). Patients who are in a pre-relapse state are hypothesized to have failing feedback systems to maintain homeostasis (Davis et al. 1991). Because indirect rather than direct dopamine agonists are more potent in worsening psychosis in schizophrenic patients, and indirect dopamine agonists also increase noradrenergic activity, we hypothesized in 1982 that noradrenergic hypersensitivity (van Kammen et al. 1982c) may be associated with the stimulant-induced worsening in psychosis seen in the presence of dopamine receptor blockade (Lieberman et al. 1984). Presumably both dopamine and NE are involved. Van Kammen et al. (1982a) reported that CSF NE correlated with the change in tension following 20 mg *d*-amphetamine, but CSF MHPG correlated significantly with the change in psychosis. In these same studies baseline CSF HVA did not correlate with the amphetamine response, but CSF DOPAC correlated with increases in activity. Therefore, van Kammen et al. (1982a) interpreted the data to indicate that NE, rather than dopamine release, played a role in the psychotogenic effects of amphetamine. This was further based on the *d*-amphetamine-induced release of norepinephrine (Coyle and Snyder 1969) and the increased CSF norepinephrine levels in schizophrenia (Lake et al. 1980; Kemali et al. 1982) that declined relative to the antipsychotic effect of pimozide (Sternberg et al. 1981). (For an additional review see Hornykiewicz 1982; van Kammen and Kelley 1991; Maas et al. 1993a,b.) The difference between the relapsing and nonrelapsed groups may explain some of the discrepancies in noradrenergic findings reported in the literature (van Kammen and Kelley 1991). In spite of all that, Innis et al. (1992) have shown that amphetamine-induced dopamine release may wash the antipsychotic off the D_2 receptors; whether NE release does the same is unknown. So at this time we propose that both dopamine and norepinephrine are setting the threshold for psychotic relapse.

Potential Interpretation of the Role of NE in Relapse

CSF NE in schizophrenia was elevated in drug-free relapsed patients, but also in those same patients prior to haloperidol withdrawal (van Kammen et al. 1989). This increase suggested to us that CSF norepinephrine was potentially a state-dependent or episode marker. More chronic and stable patients tend to have lower CSF NE values, whereas more acutely ill, relapsing patients may have higher levels. It is difficult to interpret the absolute level because of the interaction between release and receptor activity, e.g., α_2 receptor (Scheinin et al. 1991).

There are several studies (see Table 1) that reported increased (Kafka et al. 1985), decreased (Rice et al. 1984), or no difference (Bondy et al. 1984; Rice et al. 1984) in α_2 receptor binding. Clonidine challenge test (CCT) studies also have shown that schizophrenic patients have variable α_2 activity: increased (Matussek et al. 1980) and unchanged GH response compared with normal controls (Lal et al. 1983). We found that most of our medicated patients had lower than normal GH responses to the CCT, but that those who went on to relapse had the relatively higher values in GH response than the nonrelapsers. The conceptual difficulty is that the elevated GH responses to clonidine in the relapsers were still decreased or within the normal range. Functional tests such as with clonidine challenges are probably preferable, but they increase the attrition rate and complexity of doing these studies in schizophrenic patients. Pandey et al. (1989) reported that higher platelet binding was associated with a better antipsychotic drug response. We found that higher [^{3}H]-clonidine binding during chronic haloperidol treatment was associated with relapse following haloperidol withdrawal within 6 weeks (Table 2).

These studies were designed to obtain measures in patients chronically treated with haloperidol, in whom haloperidol was replaced over night with placebo for 6 weeks. We found in addition to the dose of haloperidol, that the BPRS paranoia cluster was significantly higher, while the BPRS anxiety item was significantly lower in those who went on to relapse (van Kammen et al. 1989). We then examined CSF NE, MHPG, and chromogranin A, the dose of haloperidol, the BPRS paranoia cluster and the anxiety item in 50 haloperidol-treated patients prior to drug withdrawal and did a logistic regression to discriminate between those who relapsed within 6 weeks and those who did not. We found that we were able to predict 86% of the patients correctly using these variables. Previously, Andrews et al. (1976) had pointed out that drug use should be controlled for in relapse studies (dose and type of drug) as well as other treatments, compliance before drug withdrawal, and drug-free time. In Table 2 we summarize our published findings on drug and relapse effects.

We subsequently evaluated the CSF NE, MHPG, HVA and 5-HIAA levels as well as behavioral measures such as psychosis and anxiety in 88 patients prior to haloperidol withdrawal, 54 of whom relapsed within 6 weeks on placebo. After comparing several models using receiver operating characteristic (ROC) analysis, we used improvement χ^2 measures to determine the

Table 1. Alpha receptor studies

Reference	Subjects	Measure	Comparison with normals
Kafka et al. (1985)	3 controls 4 schizophrenics	1. No. of receptors (Kd) 2. cAMP production 3. NE inhibition of cAMP production	Increased in SZ Decreased in SZ Not different
Kafka and van Kammen (1983)	51 controls 53 schizophrenics	1. No. of receptors (Kd) 2. cAMP production	Not different Decreased in male SZ
Kafka et al. (1980)	48 controls 54 schizophrenics	1. No. of receptors (Kd) 2. cAMP production 3. NE inhibition of cAMP production	Not different Decreased in SZ Not different
Rice et al. (1984)	10 controls 10 schizophrenics	1. No. of receptors (Kd) 2. No. of binding sites (Bmax) (^{3}H-yohimbine)	Increased in SZ Decreased in SZ
Rotrosen et al. (1978)	21 controls 28 schizophrenics (18 medicated)	1. cAMP production	Decreased in SZ
Kanof et al. (1988)	31 controls 51 schizophrenics 38 depressives	1. NE inhibition of cAMP production	Not different when reduced cAMP production in SZ taken into account
Ko et al. (1986)	autopsy brains: 6 control 5 schizophrenic	1. In vitro receptor autoradiography-mean concentration of ligand bound	Decreased in SZ
Response to clonidine challenge			
Lal et al. (1983)	18 controls 13 schizophrenics	1. clonidine infusion-growth hormone response	Not different
Matussek et al. (1980)	32 controls 8 schizophrenics 10 schizoaffectives	1. clonidine infusion-growth hormone response	Response higher in schizoaffectives but not schizophrenics
Sternberg et al. (1982)	11 controls 11 schizophrenics	1. clonidine administration-plasma MHPG response	No decrease in SZ decrease in normals
van Kammen et al. (1990)	32 schizophrenics	1. clonidine challenge test-GH response	Relapsers had significantly higher GH response than nonrelapsers prior to haloperidol withdrawal

SZ, Schizophrenics

Table 2. Relapse predictors (episode markers)

	Relapsers	Nonrelapsers	p
van Kammen et al. (1989) (n = 32)			
CSF NE	.79 ± .356	.52 ± .348	.04
BPRS paranoia	7.0 ± 2.08	4.9 ± 2.03	.007
BPRS anxiety	2.1 ± .83	2.8 ± 1.15	.045
BPRS tension	2.6 ± .85	3.2 ± .86	.06
van Kammen et al. (1990) (n = 32)			
CCT GH 90 min	7.2 ± 10.01	1.6 ± 1.46	<.05
CCT GH 120 min	5.2 ± 6.3	1.9 ± 2.09	<.035
Steinhauer et al. (1992) (n = 19)	Papillary constriction and redilation		
Constriction latency (ms)	888.9 (99.1)	1036.0 (186.6)	.05
50% Redilation (ms)	1716.3 (162.9)	1874.0 (128.3)	.03
van Kammen et al. (unpublished data)			
Haloperidol $\alpha2$ receptor binding (with haloperidol dose used as covariate)	101.6 (41.0)	73.7 (27.0)	.02
State-dependent differences:			
Peters et al. (1990) (n = 31 pairs)	Neuropeptide Y		
	Relapsers	Nonrelapsers	
Haloperidol	143 ± 25.7	135 ± 22.0	
Drug-free	149 ± 24.9	152 ± 24.9	
p-value	NS	.003	
van Kammen et al. (1992) (n = 33 pairs)	Chromogranin A		
	Relapsers	Nonrelapsers	
Haloperidol	41 ± 13.0	45 ± 25.8	
Drug-free	44 ± 15.7	61 ± 29.6	
p-value	NS	.04	

most significant factors in the model. The final model predicting relapse was calculated using the exponent of the logic function as *0.1131 + 0.0087 (CSF HVA) − 0.0192 (CSF 5-HIAA) − 0.3405 (anxiety) + 0.7484 (CSF NE) + 0.3468 (psychosis)*, and the probability of relapse (predicted values) was calculated as $1/1+e^{-(\mathrm{exp})}$.

Discussion

The failure to identify meaningful etiological subgroups through monoaminergic measures focuses attention on some of the inherent assumptions and limitations of cross-sectional research in schizophrenia (Tsuang et al. 1990; Mortimer 1992).

Among these unspoken assumptions rests the belief that the increase or decrease in a biochemical variable in a particular body fluid is the critical expression of a potentially disturbed system, and that clinically identical presentations suggest similar etiologies. Unless we clarify how our particular measure relates to the primary or secondary etiology, we may be stuck trying to answer questions that cannot be answered by our methodology.

Some proposed etiological variables of the past may be involved in the form and course of the illness without causing the illness, i.e., illness modifiers (van Kammen 1991). Methodological issues need to be rigorously addressed and communicated. There are also statistical reasons for the failure to explain variability through subgroups. The lack of bimodality in biological variables is the major stumbling block to identifying biochemical subgroups. If we are to come up with biochemical etiological subgroups, we will have to be able to define one subgroup first.

Furthermore, multiple methodologies need to be integrated. By combining different findings in the same patient populations (e.g., cognitive changes, negative symptoms, poor premorbid functioning, functional and structural changes in the brain) that are related but seem to be independent (van Kammen et al. 1986b, 1988) we may be able to increase our understanding of how they relate to each other and to the different aspects or domains of schizophrenia. This could lead to more meaningful subdivisions that are otherwise not possible. We have started to make an attempt at this by integrating studies of monoamines and peptides in CSF, plasma, and urine, in combination with cognitive studies, brain imaging, sleep, psychophysiology, and membrane studies, in addition to extensive clinical and demographic descriptions.

It is conceivable that the neurodevelopmental hypotheses of schizophrenia (Murray et al. 1992; Weinberger 1987) may better explain the variability and heterogeneity in schizophrenia, leading us more towards the difficult-to-test multifactorial models (Tsuang et al. 1990; McGuffin et al. 1987) rather than to a one-neurotransmitter, one-gene etiological model. Contenders for the primary etiology of such a multifactorial model may be found among abnormalities in membranes, nerve growth factors, cytoskeleton structure, or nCAMs, causing erroneous cell migration (Akbarian et al. 1993; Kovelman and Scheibel 1984; Jakob and Beckman 1986; Benes 1991) and neuronal connections, which may lead to changes in monoamine systems. Once we get out of the dopamine etiology mode, new research strategies for the etiology of schizophrenia may emerge. So far, we are dealing with a continuum model, not with different groups.

Heterogeneity in biochemistry has definite clinical implications in schizophrenia. The state-dependency model lets us interpret differences as they relate to behavior or to impending relapse. There is no evidence that early relapsers have a different illness from those who do not relapse as quickly, i.e., within 6 weeks. We propose that changes in serotonin and NE systems interfere with the DA system, leading to psychotic relapse. The results suggest that the feedback systems (to allow for adaptation and flexibility in behavior within "normal" functioning) are impaired in schizophrenia with an increase in stress responsiveness (van Kammen and van Kammen 1984; Albus et al. 1982; Breier et al. 1989), leading to shifts in psychotic behavior or negative symptoms (Strauss et al. 1985). Conceivably, dopamine receptor blockade masks the behavioral expression of increased stress sensitivity (e.g., increased NE activity), which without medication may be exposed in the relapsers. Antipsychotic drugs may allow patients to make use of stress reduction and of a structured environment to improve control over their lives up to a certain level of functioning.

A potential area for exploration of biochemical heterogeneity is the recent clinical division of schizophrenia into categories or domains (Carpenter 1989), which may give us the needed subgroups, such as "Kraepelinian" schizophrenia (Keefe et al. 1987), the deficit syndrome (Carpenter et al. 1988; Kelley et al. 1992), or the three (reality distortions, psychomotor poverty, and disorganization) (Bilder et al. 1985; Liddle and Barnes 1990) or five syndromes (Kay and Sevy 1990). The search for schizophrenia-related genetic changes may include such explorations of how certain genes or their alleles are related to different aspects of the disorder or the newly developed clinical subgroups. Such associations could validate the use of these divisions and may help prevent the loss of positive findings.

From the available monoamine measures we cannot certify etiological subgroups of schizophrenia, because our monoaminergic measures may be one or more steps removed from the primary cause (See Tsuang et al. 1990; Mortimer 1992). Drug response does not need to be etiologically or genetically linked to schizophrenia, but interacts with biochemical changes in the brain that are secondarily induced or state dependent. In drug nonresponders or late responders biochemical systems may have become less responsive, or the homeostasis is pushed towards maintaining abnormal behavior (Friedhoff 1986). The biochemical variability, however, may indicate episode markers, which assist us in relapse and drug response prediction and may direct us to examine the variable stress sensitivity leading to exacerbation of psychosis. Finding the causes of the variable stress supersensitivity is probably the key biochemical issue to be addressed in schizophrenia research today. By understanding the episodic changes in biochemistry we may learn more about the dynamics of the behavior change in schizophrenia and ultimately about the disorder itself.

Acknowledgments. Partial funding for this project was provided to Dr. van Kammen by the Highland Drive VAMC, the Department of Veterans Affairs Research and Development Service (Merit Review), the National Institute of Mental Health (MH44-841), and the Medical Research Fund of the American Defenders of Bataan and Corregidor.

References

Akbarian S, Viñuela A, Kim JJ, Potkin SG, Bunney WE Jr, Jones EG (1993) Distorted distribution of nicotinamide-adenine dinucleotide phosphate-diaphorase neurons in temporal lobe of schizophrenics implies anomalous cortical development. Arch Gen Psychiatry 50:178–187

Albus M, Ackenheil M, Engel RR, Muller F (1982) Situational reactivity of autonomic functions in schizophrenic patients. Psychiatry Res 6:361–370
Andrews P, Hall JN, Snaith RP (1976) A controlled trial of phenothiazine withdrawal in chronic schizophrenic patients. Br J Psychiatry 128:451–455
Angrist B, van Kammen DP (1984) CNS stimulants as tools in the study of schizophrenia. Trends Neurosci 7:388–390
Angrist B, Peselow E, Rotrosen J, Gershon S (1981) Relationships between responses to dopamine agonists, psychopathology, neuroleptic treatment response, and need for neuroleptic maintenance in schizophrenic subjects. In: Angrist B, Burrows BD, Lader M, Lingjaerde O, Sedvall G, Wheatley D (eds) Recent advances in neuropsychopharmacology. Permagon, New York, pp 49–54
Angrist B, Peselow E, Rubinstein M, Wolkin A, Rotrosen J (1985) Amphetamine response and relapse risk after depot neuroleptic discontinuation. Psychopharmacology 85:277–283
Benes FM (1991) Evidence for neurodevelopment disturbances in anterior cingulate cortex of post-mortem schizophrenic brain. Schizophr Res 5:187–188
Bilder RM, Mukherjee S, Rieder RO, Pandurangi AK (1985) Symptomatic and neuropsychological components of defect states. Schizophr Bull 11:409–419
Bondy B, Ackenheil M, Birzle W, Elbers R, Frohler M (1984) Catecholamines and their receptors in blood: evidence for alterations in schizophrenia. Biol Psychiatry 19:1377–1393
Bowers MB Jr, Swigar ME, Jatlow PI, Hoffman F, Giocoechea N (1986) Early neuroleptic response in psychotic men and women: correlation with plasma HVA and MHPG. Compr Psychiatry 27:181–185
Breier A (1989) Experimental approaches to human stress research: assessment of neurobiological mechanisms of stress in volunteers and psychiatric patients. Biol Psychiatry 26:438–462
Brown WA, Laughren TP (1981) Low serum prolactin and early relapse during neuroleptic treatment and withdrawal. Am J Psychiatry 136:237–239
Brown WA, Herz LR (1987) Response to neuroleptic drugs as a device for classifying schizophrenia. Schizophr Bull 15:123–129
Brown GM, Mazurek M, Allen D, Szechtman B, Cleghorn JM (1990) Dose-response profiles of plasma growth hormone and vasopressin after clonidine challenge in man. Psychiatry Res 31:311–320
Buchanan RW, Kirkpatrick B, Summerfelt A, Hanlon TE, Levine J, Carpenter WT Jr (1992) Clinical predictors of relapse following neuroleptic withdrawal. Biol Psychiatry 32:72–78
Carlsson A, Lindqvist M (1963) Effect of chlorpromazine or haloperidol on formation of 3-methoxytramine and normetanephrine in mouse brain. Acta Pharmacol Toxicol 20:140–144
Carpenter WT (1989) Dissecting schizophrenia through psychopathology, monograph. The 1989 Charles Burlingame Award Lecture. Institute of Living, Hartford, CT
Carpenter WT Jr, Heinrichs DW, Wagman AMI (1988) Deficit and nondeficit forms of schizophrenia: the concept. Am J Psychiatry 145:578–583
Chang W-H, Chen T-Y, Lin S-K, Lung F-W, Lin W-L, Hu W-H, Yeh E-K (1990) Plasma catecholamine metabolites in schizophrenics: Evidence for the two-subtype concept. Biol Psychiatry 27:510–518
Cleghorn JM, Brown GM, Brown PJ, Kaplan RD, Dermer SW, MacCrimmon DJ, Mitton J (1983) Growth hormone responses to apomorphine HCl in schizophrenic patients on drug holidays and at relapse. Br J Psychiatry 142:482–488
Condray R, Steinhauer SR (1992) Schizotypal personality disorder in individuals with and without schizophrenic relatives: similarities and contrasts in neurocognitive and clinical functioning. Schizophr Res 7:33–41
Coyle JT, Snyder SH (1969) Catecholamine uptake by synaptosomes in homogenates of rat brain. Stereospecificity in different areas. J Pharmacol Exp Ther 170:221–231

Crawley JCW, Owens DGC, Crow TJ, Poulter M, Johnstone EC, Smith T, Oldland SRD, Veall N, Owen F, Zanelli GD (1986) Dopamine D2 receptors in schizophrenia studied in vivo. Lancet 2:224–225

Crow TJ, Owen F, Cross AJ, Johnstone EC, Joseph MH, Longden A (1980) The dopamine receptor as the site of the primary disturbance in the type-I syndrome of schizophrenia. In: Usdin E, Sourkes Tl, Youdim MBH (eds) Enzymes and neurotransmitters in mental disease. Wiley, New York, pp 559–572

Davidson M, Davis KL (1988) A comparison of plasma homovanillic acid concentrations in schizophrenic patients and normal controls. Arch Gen Psychiatry 45:561–563

Davidson M, Keefe RSE, Mohs RC, Siever LJ, Losonczy MF, Horvath TB, Davis KL (1987) L-dopa challenge and relapse in schizophrenia. Am J Psychiatry 144:934–938

Davila R, Manero E, Zumarraga M, Andia I, Schweitzer JW, Friedhoff AJ (1988) Plasma homovanillic acid as a predictor of response to neuroleptics. Arch Gen Psychiatry 45:561–563

Davis KL, Davidson M, Mohs RC, Kendler KS, Davis BM, Johns CA, DeNigris Y, Horvath TB (1985) Plasma homovanillic acid concentration and the severity of schizophrenic illness. Science 227:1601–1602

Davis KL, Kahn RS, Ko G, Davidson M (1991) Dopamine in schizophrenia: a review and reconceptualization. Am J Psychiatry 148:1474–1486

Dencker SJ, Malm U, Lepp M (1986) Schizophrenic relapse after drug withdrawal is predictable. Acta Psychiatr Scand 73:181–185

Delisi LE, Dauphinais D (1989) Neuroleptic responsiveness in siblings concordant for schizophrenia. Arch Gen Psychiatry 46:477

Docherty JP, van Kammen DP, Siris SG, Marder SR (1978) Stages of onset of schizophrenic psychosis. Am J Psychiatry 135:420–426

Farde L, Wiesel F, Stone-Elander S, Halldin C, Nordstrom A, Hall H, Sedvall G (1990) D2 dopamine receptors in neuroleptic-naive schizophrenic patients. Arch Gen Psychiatry 47:213–219

Friedhoff AJ (1986) A dopamine-dependent resitutive system for the maintenance of mental normalcy. Ann NY Acad Sci 463:47–52

Garver DL, Kelly K, Fried KA, Magnusson M, Hirschowitz J (1988) Drug response patterns as a basis of nosology for the mood-incongruent psychoses (the schizophrenias). Psychol Med 18:1–13

Gattaz WF, Riederer P, Reynolds GP, Gattaz D, Beckmann H (1983) Dopamine and noradrenaline in the cerebrospinal fluid of schizophrenic patients. Psychiatry Res 8:243–250

Gillin JC, Kaplan J, Stillman R, Wyatt RJ (1976) The psychedelic model of schizophrenia: the case of N,N-dimethyltryptamine. Am J Psychiatry 133:203–214

Green AI, Faraone SV, Brown WA (1990) Prolactin shifts after neuroleptic withdrawal. Psychiatry Res 32:213–219

Heinrichs DW, Carpenter WT Jr (1985) Prospective study of prodromal symptoms in schizophrenic relapse. Am J Psychiatry 142:371–373

Henn FA, Henn SW (1982) Phospholipids as markers for schizophrenia. In: Usdin E, Hanin I (eds) Biological markers in psychiatry and neurology. Permagon New York, pp 183–185

Herz MI, Szymanski HV, Simon JC (1982) Intermittent medication for stable schizophrenic outpatients: an alternative to maintenance medication. Am J Psychiatry 139:917–922

Herz MI, Glazer WM, Mostert A, Sheard MA, Szymanski HV, Hafez H, Mirza M, Vana J (1991) Intermittent vs. maintenance medication in schizophrenia. Arch Gen Psychiatry 48:333–339

Hirsch SR, Jolley AG, Morrison E, McRink A, Wilson L (1990) Trial of brief intermittent neuroleptic prophylaxis for selected schizophrenic outpatients: clinical and social outcome at 2 years. Schizophr Res 3:40

Holzman PS, Kringlen E, Matthysse S, Flanagan SD, Lipton RB, Cramer G, Levin S, Lange K, Levy DL (1988) A single dominant gene can account for eye tracking dysfunctions and schizophrenia in offspring of discordant twins. Arch Gen Psychiatry 45:641–647
Hornykiewicz O (1982) Brain catecholamine in schizophrenia – a good case for noradrenaline. Nature 299:484–486
Innis RB, Malison RT, Al-Tikriti M, Hoffer PB, Sybirska EH, Seibyl JP, Zoghbi SS, Baldwin RM, Laruelle M, Smith EO, Charney DS, Heninger G, Elsworth JD, Roth RH (1992) Amphetamine-stimulated dopamine release competes in vivo for [^{123}I]IBZM binding to the D2 receptor in nonhuman primates. Synapse 10:177–184
Jakob H, Beckmann H (1986) Prenatal developmental disturbances in the limbic allocortex in schizophrenics. J Neural Transm 65:303–326
Janowsky DS, Davis JM (1976) Methylphenidate, dextroamphetamine, and levamphetamine: effects on schizophrenic symptoms. Arch Gen Psychiatry 33:304–308
Janowsky DS, El-Yousef K, Davis JM, Sekerke HJ (1973) Provocation of schizophrenic symptoms by intravenous administration of methylphenidate. Arch Gen Psychiatry 28:185–191
Johnson DAW (1976) The expectation of outcome from maintenance therapy in chronic schizophrenic patients. Br J Psychiatry 128:246–250
Johnson DAW, Pasterski G, Ludlow JM, Street K, Taylor RDW (1983) The discontinuance of maintenance neuroleptic therapy in chronic schizophrenic patients: drug and social consequences. Acta Psychiatr Scand 67:339–352
Johnstone EC, Owens DGC, Frith CD, Crow TJ (1986) The relative stability of positive and negative features in chronic schizophrenia. Br J Psychiatry 150:60–64
Kafka MS, van Kammen DP, Kleinman JE (1980) α-Adrenergic receptor function in schizophrenia. Psychopharmacol Bull 16:91–94
Kafka MS, van Kammen DP (1983) α-Adrenergic receptor function in schizophrenia. Arch Gen Psychiatry 40:264–270
Kafka MS, Siever LJ, Nurnberger JI, Uhde TW, Targum S. Cooper DMJ, van Kammen DP, Tokola NS (1985) Platelet alpha-adrenergic receptor function in affective disorders and schizophrenia. Psychopharmacol Bull 21:599–602
Kane J, Honigfeld G, Singer J, Meltzer H (1988) Clozapine for the treatment-resistant schizophrenic. Arch Gen Psychiatry 45:789–796
Kanof PD, Johns CA, Davidson M, Siever LJ, Coccaro EF, Davis KL (1988) Platelet alpha$_2$-adrenergic receptor function in psychiatric disorders. Psychiatry Res 23: 11–22
Karoum F, Karson CN, Bigelow LB, Lawson WB, Wyatt RJ (1987) Preliminary evidence of reduced combined output of dopamine and its metabolites in chronic schizophrenia. Arch Gen Psychiatry 44:604–607
Kay SR, Sevy S (1990) Pyramidical model of schizophrenia. Schizophr Bull 16:537–545
Keefe RSE, Mohs RC, Losonczy MF, Davidson M, Silverman JM, Kendler KS, Horvath TB, Nora R, Davis KL (1987) Characteristics of very poor outcome schizophrenia. Am J Psychiatry 144:889–895
Kelley ME, Gilbertson M, Mouton A, van Kammen DP (1992) Deterioration in premorbid functioning in schizophrenia: a developmental model of negative symptoms in drug-free patients. Am J Psychiatry 149:1543–1548
Kemali D, Delvecchio M, Maj M (1982) Increased noradrenaline levels in CSF and plasma of schizophrenic patients. Biol Psychiatry 17:711–717
Kirkpatrick B, Carpenter WT, Maeda K, Buchanan RW, Breier A, Tamminga CA (1992) Plasma prolactin as a predictor of relapse in drug-free schizophrenic outpatients. Biol Psychiatry 32:1049–1054
Ko GN, Unnerstall JR, Kuhar MJ, Wyatt RJ, Kleinman JE (1986) Alpha-2 adrenergic agonist binding in schizophrenic brains. Psychopharmacology Bull 22:1011–1016
Ko GN, Jimerson DC, Wyatt RJ, Bigelow LB (1988) Plasma 3-methoxy-4-hydroxyphenylglycol changes associated with clinical state and schizophrenic subtype. Arch Gen Psychiatry 45:842–846

Kovelman JA, Scheibel AB (1984) A neurohistological correlate of schizophrenia. Biol Psychiatry 19:1601–1621
Lake CR, Sternberg DE, van Kammen DP, Ballenger JC, Ziegler MG, Post RM, Kopin IJ, Bunney WE (1980) Schizophrenia: elevated cerebrospinal fluid norepinephrine. Science 207:331–333
Lal S, Nair NPV, Thavundayil JX, Monks RC, Gudya H (1983) Clonidine-induced growth hormone secretion in chronic schizophrenia. Acta Psychiatr Scand 68:82–88
Lambert GW, Eisenhofer G, Cox HS, Horne M (1991) Direct determination of homovanillic acid release from the human brain, an indicator of central dopaminergic activity. Life Sci 49:1061–1072
Liddle PF, Barnes TRE (1990) Syndromes of chronic schizophrenia. Br J Psychiatry 157:558–561
Lieberman JA, Kane JM, Gadaleta D, Brenner R, Lesser MS, Kinon B (1984) Methylphenidate challenge as a predictor of relapse in schizophrenia. Am J Psychiatry 141:633–638
Lieberman JA, Jody D, Geister S, Alvir J, Loebel A, Szymanski S, Woaner M, Barenstein M (1993) Time course and biological correlates of treatment response in first-episode schizophrema. Arch Gen Psychiatry 50:369–376
Loebel AD, Lieberman JA, Alvir JM, Mayerhoff DI (1992) Duration of psychosis and outcome in first-episode schizophrenia. Am J Psychiatry 149:1183,1188
Maas JW, Contreras SA, Miller AL, Berman N, Bowden CL, Javors MA, Seleshi E, Weintraub S (1993a) Studies of catecholamine metabolism in schizophrenia/psychosis-I. Neuropsychopharmacology 8:97–109
Maas JW, Contreras SA, Miller AL, Berman N, Bowden CL, Javors MA, Seleshi E, Weintraub S (1993b) Studies of catecholamine metabolism in schizophrenia/psychosis-II. Neuropsychopharmacology 8:111–116
MacKay AV (1980) Positive and negative schizophrenic symptoms and the role of dopamine. Br J Psychiatry 137:379–383
Marder SR, van Kammen DP, Bunney WE Jr (1979) Prediction of drug-free improvement in schizophrenic psychosis. Arch Gen Psychiatry 36:1080–1085
Matussek N, Ackenheil M, Hippius H, Müller F, Schröder H-Th, Schultes H, Wasilewski B (1980) Effect of clonidine on growth hormone release in psychiatric patients and controls. Psychiatr Res 2:25–36
Mazure CM, Nelson JC, Jatlow PI, Bowers MB (1991) Plasma free homovanillic acid (HVA) as a predictor of clinical response in acute psychosis. Biol Psychiatry 30:475–482
McGuffin P, Farmer A, Gottesman II (1987) Is there really a split in schizophrenia? The genetic evidence. Br J Psychiatry 150:581–592
McKusick VA (1969) On lumpers and splitters, or the nosology of genetic disease. Perspect Biol Med 12:298–301
Moises HW, Gelernter J, Giuffra LA, Zarcone VP (1991) No linkage between D2 dopamine receptor gene region and schizophrenia. Arch Gen Psychiatry 48:643–647
Mortimer AM (1992) Phenomenology, its place in schizophrenia research. Br J Psychiatry 161:293–297
Müller-Spahn F, Ackenheil M, Bondy B, May G, Rüther E (1985) Growth hormone response to graded doses of apomorphine HCl in normals and schizophrenic patients: relation to psychotic decompensation? In: Shagass C, Josiassen RC, Bridger WH, Weiss KJ, Stoff D, Simpson GM (eds) Biological psychiatry 1985. Elsevier Science, New York, pp 1074–1076
Murray RM, O'Callaghan E, Castle DJ, Lewis SW (1992) A neurodevelopmental approach to the classification of schizophrenia. Schizophr Bull 18:319–332
Neuchterlein KH, Dawson ME (1984) A heuristic vulnerability/stress model of schizophrenic episodes. Schizophr Bull 10:300–312
Neuchterlein KH, Dawson ME, Gitlin M, Ventura J, Goldstein MJ, Snyder KS, Yee CM, Mintz J (1992) Developmental processes in schizophrenic disorders: Longitudinal studies of vulnerability and stress. Schizophr Bull 18:387–425

Öhlund LS, Lindström LH, Öhman A (1992) Electrodermal orienting response and central nervous system dopamine and serotonin activity in schizophrenia. J Nerv Ment Dis 180:304–313
Pandey GN, Janicak PG, Javaid JI, Davis JM (1989) Increased ^{3}H-clonidine binding in the platelets of patients with depressive and schizophrenic disorders. Psychiatry Res 28:83–88
Parnas J, Cannon TD, Jacobsen B, Schulsinger H, Schulsinger F, Mednick SA (1993) Lifetime DSM-III-R diagnostic outcomes in the offspring of schizophrenic mothers. Results from the Copenhagen high-risk study. Arch Gen Psychiatry 50:707–714
Penrose LS (1952) Measurement of pleiotropic effects in phenyl ketonuria. Ann Eugenics 18:120–124
Peters JL, van Kammen DP, Gelernter J, Yao JK, Shaw D (1990) Neuropeptide Y-like immunoreactivity in schizophrenia: relationships with clinical measures. Schizophr Res 3:287–294
Pickar D, Breier A, Hsiao JK, Doran AR, Wolkowitz OM, Pato CN, Konicki E, Potter WZ (1990) Cerebrospinal fluid and plasma monoamine metabolites and their relation to psychosis. Implications for regional brain dysfunction in schizophrenia. Arch Gen Psychiatry 47:641–648
Post RM, Fink E, Carpenter WT, Goodwin FK (1975) Cerebrospinal fluid amine metabolites in acute schizophrenia. Arch Gen Psychiatry 32:1063–1069
Post RM, Gold PW, Rubinow DR, Bunney WE Jr, Ballenger JC, Goodwin FK (1983) Cerebrospinal fluid as neuroregulatory pathway. Peptides in neuropsychiatric illness. In: Wood JH (ed) Neurobiology of cerebrospinal fluid, 2. Plenum New York, pp 107–141
Rapaport MH, McAllister CG, Pickar D, Nelson DL, Paul SM (1989) Elevated levels of soluble interleukin 2 receptors in schizophrenia (letter). Arch Gen Psychiatry 46: 291–292
Reveley MA, Reveley AM, Clifford CA, Murray RM (1983) Genetics of platelet MAO activity in discordant schizophrenic and normal twins. Br J Psychiatry 142:560–565
Rice HE, Smith CB, Silk KR, Rosen J (1984) Platelet alpha2-adrenergic receptors in schizophrenic patients before and after phenothiozine treatment. Psychiatry Res 12:69–77
Roberts E (1973) An hypothesis suggesting that there is a defect in the GABA system in schizophrenia. An essay. Neurosci Res Prog Bull 10:468–482
Rotrosen J, Miller AD, Mandio D, Traficante LJ, Gershon S (1978) Reduced PGE_1 stimulated ^{3}H-cAMP accumulation in platelets from schizophrenics. Life Sci 23: 1989–1996
Scheinin M, Karhuvaara S, Ojala-Karlsson P, Kallio A, Koulu M (1991) Plasma 3,4-dihydroxyphenylglycol (DHPG) and 3-methoxy-4-hydroxyphenylglycol (MHPG) are insensitive indicators of α_2-adrenoceptor mediated regulation of norepinephrine release in healthy human volunteers. Life Sci 49:75–84
Seeman P, Ulpian C, Bergeron C, Riederer P, Jellinek K, Gabriel E, Reynolds GP, Tourtelotte WW (1984) Bimodal distribution of dopamine receptor densities in brains of schizophrenics. Science 225:728–731
Siever LJ, Silverman JM, Horvath TB, Klar H, Coccaro E, Keefe RSE, Pinkham L, Rinaldi P, Mohs RC, Davis KL (1990) Increased morbid risk for schizophrenia-related disorders in relatives of schizotypal personality disordered patients. Arch Gen Psychiatry 47:634–640
Slater E (1947) Genetic causes of schizophrenic symptoms. Monatsschr Psychiatrie Neurol 113:50–58 (Reprinted in Man, mind and heredity, selected papers of Eliot Slater on psychiatry and genetics. Johns Hopkins Press, Baltimore, 1971)
Stein L, Wise CD (1971) Possible etiology of schizophrenia: progressive damage to the noradrenergic reward system by 6-hydroxydopamine. Science 171:1032–1036
Steinhauer SR, van Kammen DP, Colbert K, Peters JL, Zubin J (1992) Pupillary constriction during haloperidol treatment as a predictor of relapse following drug withdrawal in schizophrenic patients. Psychiatry Res 43:287–298

Sternberg DE, van Kammen DP, Lake CR, Ballenger JC, Marder SR, Bunney WE Jr (1981) The effect of pimozide on CSF norepinephrine in schizophrenia. Am J Psychiatry 138:1045–1051
Sternberg DE, Charney DS, Heninger GR, Leckman JF, Hafstad KM, Landis DH (1982) Impaired presynaptic regulation of norepinephrine in schizophrenia. Effects of clonidine in schizophrenic patients and normal controls. Arch Gen Psychiatry 39: 285–289
Sternberg DE, van Kammen DP, Lerner P, Ballenger JC, Marder SR, Post RM, Bunney WE (1983) CSF dopamine beta-hydroxylase in schizophrenia. Arch Gen Psychiatry 40:743–747
Strauss JS, Hafez H, Lieberman P, Harding CM (1985) The course of psychiatric disorder. III. Longitudinal principles. Am J Psychiatry 142:289–296
Subotnik KL, Nuechterlein KH (1988) Prodromal signs and symptoms of schizophrenic relapse. J Abnorm Psychol 97:405–412
Terenius L, Wahlstrom A, Lindstrom L, Widerlöv (1976) Increased CSF levels of endorphines in chronic psychosis. Neurosci Lett 3:157–162
Torrey EF, Peterson MR (1976) The viral hypothesis of schizophrenia. Schizophr Bull 2:136–146
Tsuang MT, Lyons MJ, Faraone SV (1990) Heterogeneity in schizophrenia. Conceptual models and analytic stategies. Br J Psychiatry 156:17–26
van der Velde CD (1976) Variability in schizophrenia. Reflection of a regulatory disease. Arch Gen Psychiatry 33:489–496
van Kammen DP (1977) Gamma-aminobutyric acid (GABA) and the dopamine hypothesis of schizophrenia. Am J Psychiatry 134:138–143
van Kammen DP (1991) The biochemical basis of relapse and drug response in schizophrenia: review and hypothesis. Psychol Med 21:881–895
van Kammen DP, Kelley M (1991) Dopamine and norepinephrine activity in schizophrenia: an integrative perspective. Schizophr Res 4:173–191
van Kammen DP, Sternberg DE (1980) CSF studies in schizophrenia. In: Wood JH (ed) Neurobiology of cerebrospinal fluid, vol I. Plenum, New York, pp. 719–742
van Kammen DP, van Kammen WB (1984) The amphetamine challenge test in schizophrenia: a stress response? In: Usdin E, Kvetnansky R, Axelrod J (eds) Stress: the role of catecholamines and other neurotransmitters, vol 2. Gordon & Breach Science Publishers, London, pp 957–966
van Kammen DP, Docherty JP, Bunney WE Jr (1982a) Prediction of early relapse after pimozide discontinuation by response to d-amphetamine during pimozide treatment. Biol Psychiatry 17:233–242
van Kammen DP, Docherty JP, Marder SR, Rayner JN, Bunney WE Jr (1982b) Long-term pimozide pretreatment differentially affects behavioral responses to dextroamphetamine in schizophrenia. Arch Gen Psychiatry 39:275–281
van Kammen DP, Bunney WE Jr, Docherty JP, Marder SR, Ebert MH, Rosenblatt JE, Rayner JN (1982c) D-amphetamine-induced heterogeneous changes in psychotic behavior in schizophrenia. Am J Psychiatry 139:991–997
van Kammen DP, Peters J, van Kammen WB (1986a) Cerebrospinal fluid studies of monoamine metabolism in schizophrenia. In: Roy A (ed) Schizophrenia: Psychiatric Clinics of North America. WB Saunders, Philadelphia, pp 81–97
van Kammen DP, van Kammen WB, Mann LS, Seppala T, Linnoila M (1986b) Dopamine metabolism in the cerebrospinal fluid of drug-free schizophrenic patients with and without cortical atrophy. Arch Gen Psychiatry 43:978–983
van Kammen DP, van Kammen WB, Peters JL, Goetz K, Neylan T (1988) Decreased slow-wave sleep and enlarged lateral ventricles in schizophrenia. Neuropsychopharmacology 1:265–271
van Kammen DP, Peters J, van Kammen WB, Nugent A, Goetz KL, Yao J, Linnoila M (1989) CSF norepinephrine in schizophrenia is elevated prior to relapse after haloperidol withdrawal. Biol Psychiatry 26:176–188
van Kammen DP, Peters J, Yao J, McAdam D, Mouton A, Breeding W (1990)

Prediction of relapse following neuroleptic withdrawal: the role of noradrenaline. Clin Neuropharmacology 13:436–437
van Kammen DP, Peters J, Yao J, Neylan T, Beuger M, Pontius E, O'Connor DT (1992) CSF chromogranin A-like immunoreactivity in schizophrenia:assessment of clinical and biochemical relationships. Schizophr Res 6:31–39
van Kammen DP, Peters JL, Kelley ME, Gilbertson MW, Gurklis JA, O'Connor DT (1994) CSF dopamine β-hydroxylase in schizophrenia: associations with premorbid functioning and brain CT scan measures. Am J Psychiatry (in press)
van Rossum JM (1966) The significance of dopamine-receptor blockade for the mechanism of action of neuroleptic drugs. Arch Int Pharmacodyn 160:492–494
Ventura J, Nuechterlein KH, Hardesty JP, Gitlin M (1992) Life events and schizophrenic relapse after withdrawal of medication. Br J Psychiatry 161:615–620
Verhoeven WMA, van Praag HM, van Ree JM, De Wied D (1979) Improvement of schizophrenic patients treated with [des-tyr^1]-γ-endorphin (DTγE). Arch Gen Psychiatry 36:294–298
Weinberger DR (1987) Implications of normal brain development for the pathogenesis of schizophrenia. Arch Gen Psychiatry 44:660–669
Widerlöv E (1988) A critical appraisal of CSF monoamine metabolite studies in schizophrenia. Ann NY Acad Sci 537:309–323
Wolkin A, Barouche F, Wolf AP, Rotrosen J, Fowler JS, Shiue C-Y, Cooper TB, Brodie JD (1989) Dopamine blockade and clinical response: evidence for two biological subgroups of schizophrenia. Am J Psychiatry 146:905–908
Wong DF, Wagner HN Jr, Tune LE, Dannals RF, Pearlson GD, Links JM, Tamminga CA, Broussolle EP, Ravert HT, Wilson AA, Toung JKT, Malat J, Williams JA, O'Tuama LA, Snyder SH, Kuhar MJ, Gjedde A (1986) Positron emission tomography reveals elevated D2 dopamine receptors in drug-naive schizophrenics. Science 234: 1558–1563
Yao JK, van Kammen DP (1994) Red blood cell dynamics in schizophrenia. I. Membrane fluidity. Schizophr Res 11:209–216
Zubin J, Steinhauer S (1981) How to break the logjam in schizophrenia. J Nerv Ment Dis 169:477–492

Microscopic Findings in the Cortex and Hippocampus of Schizophrenic and Schizoaffective Patients

F. M. Benes

Introduction

Postmortem studies reported over the past 10 years have demonstrated a variety of neuropathological changes, including shrinkage (Bogerts et al. 1985; Falkai and Bogerts 1986; Brown et al. 1986) and neuronal loss (Benes et al. 1986, 1991a; Falkai et al. 1988; Jeste and Lohr 1989), in several different corticolimbic regions of schizophrenics. Because there has been no convincing evidence of gliosis in the brains of schizophrenic patients (Benes et al. 1986, 1991a; Brown et al. 1986; Falkai and Bogerts 1986), a typical adult pattern of neuronal degeneration, like that seen in Alzheimer's disease, does not account for the observed loss of neurons. An alternative pathogenic mechanism that is believed to account for the reduced numbers of neurons observed in postmortem schizophrenic brain is a disturbance of normal cell migration during early brain development (Kovelman and Scheibel 1984; Jakob and Beckmann 1986; Benes et al. 1986, 1991a; Akbarian et al. 1993).

As postmortem studies focus on the possibility that schizophrenia may be a neurodevelopmental disorder, some important issues concerning the reported findings must be addressed. For example, it is not known whether the various findings reported for schizophrenics occur uniformly in all subtypes of the disorder. In other words, do all patients with this disorder show the same pattern of histopathological findings, or are there unique alterations that are associated with one subtype but not another? Although the DSM-III recognizes several subtypes of schizophrenia that include the disorganized, paranoid, catatonic, and residual types, the cohorts available for postmortem investigations typically show a preponderance of patients who can be distinguished from one another by the presence or absence of superimposed mood disturbances (Benes et al. 1991a,b) of either a depressed or manic type. This diagnostic distinction is of potential value in understanding the etiology of schizophrenia, because schizophrenic patients who show mood disturbances tend to have first-degree relatives with both schizophrenia and affective disorder (Fowler 1978; Levitt and Tsuang 1988), while those without mood disturbances primarily have either schizophrenia or the schizotypal personality in their

Laboratory for Structural Neuroscience, McLean Hospital, Belmont, MA; Department of Psychiatry and Program in Neuroscience, Harvard Medical School, Boston, MA 02115, USA

families (Kendler et al. 1985). Thus, if there is a heritable factor that is common to both disorders, it might theoretically give rise to postmortem findings that can be identified in both subtypes of schizophrenia (Benes et al. 1991a). Alternatively, if a particular microscopic finding is found exclusively in schizoaffective patients, but not in those without mood disturbances, such an alteration could theoretically be related to a "gene" or some other risk factor for mood disorder.

This paper will compare the results of several different postmortem studies of the anterior cingulate cortex, prefrontal region, and hippocampal formation of subjects who satisfy the Feighner criteria for schizophrenia (Feighner et al. 1972) and, in some cases, who also satisfy the DSM-III criteria for mania and/or depression. The discussion will consider whether certain microscopic findings (a) are selectively present in only one diagnostic subgroup of schizophrenia, (b) are present in one or several different brain regions of each subtype, and (c) show similar or dissimilar patterns of findings when several different variables have been examined in schizophrenics without superimposed mood disorder versus those with them.

Methods

Tissues

All human brain tissue was obtained through the Human Brain Tissue Resource Center, directed by Edward D. Bird at McLean Hospital. Each specimen was routinely evaluated by a neuropathologist for any evidence of Alzheimer's disease, cerebrovascular disease, or other processes that would confound the microscopic analyses. Antemortem clinical records were collected for each case and a retrospective psychiatric diagnosis was established using the criteria of Feighner et al. (1972) for schizophrenia; for those satisfying them, the DSM-III criteria for mania and/or depression were also applied to determine whether some had an affective form of schizophrenia (schizoaffective disorder). Demographic information including age (years), postmortem interval (hours), fixation interval (months), hypoxia, and neuroleptic exposure was obtained for each case.

Microscopic Analyses

The results described in this paper are derived from several different studies that have previously been published elsewhere, including cell counts in anterior cingulate and prefrontal cortices (Benes et al. 1991a), stereomorphometric analyses of the hippocampal formation (Benes et al. 1991b), autoradiographic localizations of $GABA_A$, receptor binding (Benes et al. 1992b), and immunoperoxidase staining of glutamate-containing vertical neuronal processes (Benes et al. 1992a). The specific details of each method can be found in the report cited. In addition, data from unpublished studies that either have been submitted ($GABA_A$ receptor binding in prefrontal cortex) or are being prepared for publication ($GABA_A$ and NMDA receptor binding in the hippocampal formation) are also included in this paper.

Results

Basic Stereomorphometry

Anterior Cingulate and Prefrontal Cortices

A differential cell counting study has revealed a generalized reduction in the density of nonpyramidal (small) neurons, but not pyramidal cells, in most layers of the anterior cingulate cortex (Benes et al. 1991a). Schizophrenic patients with superimposed mood disturbances ($n = 9$) showed more of a reduction (e.g., 39% vs. 19% in layer II) than those without them ($n = 9$), and the significant differences occurred primarily for the schizoaffective patients ($p < 0.02$ in layer II). In the prefrontal cortex, a reduction in the density of nonpyramidal cells was found only in layer II and, to a lesser extent, in layer I. Both the schizophrenic ($n = 8$) and schizoaffective ($n = 7$) patients showed equivalent reductions (e.g., 40% and 30%, respectively in layer II), in the density of interneurons, although the level of significance ($p = 0.002$ vs. $p = 0.05$ in layer II) was greater for the patients without superimposed mood disturbances. In layer V of this region, there was a significant increase in the density of pyramidal neurons for both subgroups of patients, but the schizoaffectives showed a larger effect size (49% vs. 38%) with a higher level of significance ($p = 0.002$ vs. $p = 0.05$). In sum, the schizoaffective patients, for the most part, showed the greatest differences in neuronal cell density in the two cortical areas examined when compared with normal controls. In both cortical regions examined, neither the schizophrenic nor the schizoaffective groups showed any difference in the density of glial cells, suggesting that a typical adult pattern of neuronal degeneration does not account for the changes in neuronal density observed in the anterior cingulate and prefrontal cortices of schizophrenic patients, either with or without superimposed mood disturbances (Benes et al. 1991a).

Hippocampal Formation

A study of the posterior hippocampus at the level of the lateral geniculate nucleus in normal and schizophrenic subjects did not show significant differences in the cross-sectional area of this region, the orientation of pyramidal neurons at the CA1-prosubicular interface, or the number of pyramidal cells in sectors CA1–4 (Benes et al. 1991b). The size of pyramidal neurons, however, was consistently smaller in the schizophrenic group. When the patients were distinguished according to the presence ($n = 7$) or absence ($n = 7$) of superimposed mood disturbances, both subtypes showed a similar decrease in pyramidal cell size in sectors CA1–4 when compared with normal controls. In addition, the number of pyramidal neurons was significantly reduced only in CA1 of schizophrenics without mood disturbances and only in CA3 of the schizoaffective patients.

Glutamate-Immunoreactive Vertical Axons

The density of glutamate-immunoreactive vertical axons was previously reported to be significantly increased in the anterior cingulate, but not the prefrontal, cortex of schizophrenics (Benes et al. 1992a). Schizophrenics, both with ($n = 11$) and without ($n = 6$) superimposed mood disorders, showed an increase (100% and 58%, respectively) in the density of these fibers, although the differences attained significance in the schizoaffective group (Table 1).

$GABA_A$ Receptor Binding Activity

Anterior Cingulate Cortex

A significant increase in the amount of $GABA_A$ receptor binding activity has been reported for neuronal cell bodies in layers II and III, but not V and VI, of the anterior cingulate cortex of schizophrenics (Benes et al. 1992b). Schizophrenics both with ($n = 3$) and without ($n = 3$) superimposed mood disturbances showed increased binding (Table 2). In layer II, the values for the latter were more than double those of controls ($p = .001$), while those for the schizoaffective subjects were only 64% higher ($p = .007$). Taking these findings together with those described above for the cell counting study, the density of

Table 1. Specific $GABA_A$ receptor binding on neurons of anterior cingulate cortex

Layer	Control ($n = 8$)	Schizophrenics: Without mood disorder ($n = 3$)	Schizophrenics: With mood disorder ($n = 3$)
II	20.2 ± 2.2	44.5 ± 2.2***	33.1 ± 3.4**
III	18.2 ± 2.2	34.0 ± 2.7***	28.4 ± 5.5*
V	18.8 ± 2.0	23.5 ± 3.1	18.0 ± 2.6
VI	10.5 ± 1.0	7.6 ± 3.6	12.5 ± 2.0

The data are expressed as the number of autoradiographic grains/neuron.
***$p = .001–.002$
**$p = .007$
*$p = .03$

Table 2. Specific $GABA_A$ receptor binding in prefrontal cortex

Layer	Control (n = 16)	Schizophrenics: Without mood disorder (n = 4)	Schizophrenics: With mood disorder (n = 3)
II	8.9 ± .9	10.8 ± 1.6	19.1 ± 1.4***
III	11.4 ± 1.3	12.1 ± 1.6	23.0 ± 1.1***
V	8.9 ± 1.2	11.2 ± 1.0	15.9 ± 1.5**
VI	4.4 ± .7	5.5 ± 1.2	7.6 ± 2.2*

The data are expressed as the number of autoradiographic grains/neuron.
***p = .001
**p = .01
*p = .05

Table 3. Density of glutamate-immunoreactive vertical axons

	Control (n = 16)	Schizophrenics: Without mood disorder (n = 6)	Schizophrenics: With mood disorder (n = 11)
Ant. Cingulate Cx	2.6 ± .5	4.1 ± .7	5.2 ± .7**
Prefrontal Cx	5.9 ± .7	6.8 ± 1.3	7.2 ± .7

**p = .001

interneurons and $GABA_A$ receptor binding activity on neuronal cell bodies were found to be altered in schizophrenics both with and without mood disturbances; however, the differences for the cell counts were greater in the schizoaffective group, while those for the receptor binding were greater in the schizophrenics without mood disorders.

Prefrontal Cortex

$GABA_A$ receptor binding activity is significantly increased on neuronal cell bodies of layers II and III of the prefrontal area of schizophrenics; however, the changes noted in this region, unlike those in the anterior cingulate (see above), occurred only for the schizoaffective patients (Table 3). When the receptor binding activity was expressed separately for pyramidal neurons, the schizoaffectives showed increased binding in layers II, III, V, and VI. In layer II it was increased by 138% relative to controls (p = .001). The schizophrenics without mood disturbances showed no difference of $GABA_A$ receptor binding on pyramidal neurons of the prefrontal area.

Table 4. Specific $GABA_A$ receptor binding in CA3 sector of hippocampus

	Control (n = 15)	Schizophrenics	
		Without mood disorder (n = 4)	With mood disorder (n = 5)
Stratum oriens	15.4 ± 3.6	14.6 ± 7.4	44.1 ± 10.8***
Stratum pyramidale	24.6 ± 3.7	23.4 ± 14.9	62.2 ± 17.6
Stratum radiatum	22.2 ± 3.5	21.1 ± 13.7	53.9 ± 12.8

Data shown are the optical densities expressed as an average ±SEM for the mean value for each case within the respective groups.
***p = .002

Hippocampal Formation

Preliminary evidence indicates that there is a marked upregulation of $GABA_A$ receptor binding in the hippocampus of schizophrenic subjects compared with normal controls (Table 4). When the data are broken down according to various subregions and laminae of this brain area, the most marked changes (100–200%) were observed in the CA3 sector of the schizoaffective subjects, but there was no difference in schizophrenics without mood disturbances. In the CA1 sector, there were no differences in schizophrenics, either with or without superimposed mood disorders (not shown).

NMDA Receptor Binding Activity

In the hippocampus, NMDA-sensitive glutamate receptor binding activity showed a significant reduction in the CA1 sector of schizophrenics without superimposed mood disorders, but not in the schizoaffective subjects (Table 5).

Discussion

Prefrontal and Anterior Cingulate Cortices

The above results indicate that there are both similarities and differences in the distribution of microscopic changes in corticolimbic regions of schizophrenic patients with and without superimposed mood disturbances. Moreover, the pattern of the findings observed seems to vary according to the brain region under consideration. In the anterior cingulate and prefrontal cortices, both schizophrenic subtypes showed a decreased density of interneurons; however,

Table 5. NMDA receptor binding in CA1 sector of hippocampus

	Control (n = 15)	Schizophrenics	
		Without mood disorder (n = 4)	With mood disorder (n = 4)
Stratum oriens	62.4 ± 4.5	37.4 ± 18.8*	69.1 ± 8.7
Stratum pyramidale	128.5 ± 7.7	75.7 ± 23.9**	145.1 ± 7.1
Stratum radiatum	93.2 ± 7.1	58.9 ± 27.2*	113.5 ± 15.4

Data shown are the optical densities expressed as an average ±SEM for the mean value for each case within the respective groups.
**p = .006
*p = .03–.04

in the prefrontal area, only the schizoaffective patients showed an increase of $GABA_A$ receptor binding activity on pyramidal neurons. These findings suggest that the observed decrease of interneurons is not selectively localized to one or the other brain region, nor is it confined to only one subtype of schizophrenia. Previously, it was hypothesized that the decreased density of interneurons and increased $GABA_A$ receptor binding may be related to one another; i.e., a loss of inhibitory neurons was thought to result in a compensatory upregulation of $GABA_A$ receptors. In the absence of increased $GABA_A$ binding in the prefrontal area of schizophrenics where diminished interneuronal numbers were observed, this hypothesis may be incorrect. Similarly, the fact that the density of interneurons in the anterior cingulate region was more markedly reduced in schizoaffective patients, while $GABA_A$ receptor binding showed a greater increase in schizophrenics without mood disturbances, might further suggest that these two changes do not co-vary with one another and are not, therefore, related. It is important to point out, however, that the number of cases for the receptor binding data in each diagnostic subtype is quite small. Although the differences are quite robust and a high degree of significance was attained, future studies with a larger number of subjects will be needed to evaluate whether there is a relationship between the extent to which interneurons are reduced and $GABA_A$ binding is increased in the two schizophrenic subtypes examined here.

Hippocampal Formation

The data reported here indicate that alterations observed in the hippocampus may also show both similarities and differences for the two schizophrenic subtypes examined. A variable which showed similar changes is the size of pyramidal neurons in sectors CA1–4, which was reduced in schizophrenics both with and without superimposed mood disturbances.

Unlike the size of hippocampal pyramidal cells, two other parameters, i.e., the number of pyramidal neurons and $GABA_A$ receptor binding activity, showed some striking differences in the two subgroups. For example, the number of pyramidal neurons was significantly reduced in the CA1 sector of nonaffective schizophrenics, but this parameter was normal in the CA1 sector of schizoaffective patients. The latter subtype of schizophrenia, however, showed a significant reduction in the number of pyramidal cells and increased $GABA_A$ receptor binding activity in the CA3 sector of the hippocampus. These parameters were both relatively normal in this subregion of schizophrenics without mood disturbances. In contrast, NMDA binding activity was decreased in the CA1 sector of schizophrenics without mood disturbances, but not in the schizoaffective group. Thus, the reduction in the number of pyramidal cells observed in the CA1 sector of schizophrenics without mood disturbance seems to co-vary with a downregulation of the NMDA receptor. These combined results with respect to the two subtypes of schizophrenia discussed here might be consistent with a model in which different pathophysiological mechanisms might be associated with the loss of pyramidal neurons observed in each. Could the reduced NMDA binding in CA1 of schizophrenics without mood disorders be due to a primary increase of glutamatergic afferent activity in this region that results in excitotoxic injury to pyramidal neurons? In contrast, could the reduction in the number of pyramidal neurons in the CA3 sector of schizoaffective subjects be triggered primarily by a loss of GABAergic activity that secondarily results in an excitotoxic injury of glutamatergic afferent activity in this subregion? An alternative possibility to explain these differential changes in CA1 sector of schizophrenics without mood disorders vs. CA3 sector of schizoaffectives might be neurodevelopmental disturbances occurring at different stages of hippocampal development (Angevine 1965).

Conclusions

Overall, these results suggest that both generalized and regionally specific patterns of microscopic changes may be present in the brains of patients with two different subtypes of schizophrenia, distinguished by the presence or absence of superimposed mood disturbances. While there are differences in the patterns observed in the hippocampal formation and, to a lesser extent, the prefrontal area, the changes observed in the anterior cingulate area were similar for the two schizophrenic subtypes examined here. Thus, these preliminary findings are not consistent with the idea that schizophrenia with and schizophrenia without superimposed mood disturbances are distinct entities; rather, they seem to be part of a psychotic continuum that may involve common mechanisms in overlapping corticolimbic regions. To fully understand the implications of these findings, it will be important to determine whether the changes noted are primarily related to the pathophysiology of schizophrenia or whether they are secondary in nature. Based on the findings discussed here, it seems unlikely that any of the abnormalities thus far observed are related to a "gene" for affective disorder, because none has been found to be unique to schizophrenics who have superimposed mood disturbances. With regard to alterations that are found in both subtypes of schizophrenia, such as increased $GABA_A$ receptor binding and increased glutamatergic vertical axons in anterior cingulate cortex, it is possible that changes may be related to a gene for schizophrenia and/or the schizotypal personality. Over the coming years, systematic identification of the nature and distribution of microscopic abnormalities in corticolimbic regions of patients with different subtypes of schizophrenia may potentially provide important insights as to how the risk factors for psychosis may contribute to the primary pathophysiology of this disorder.

Acknowledgments. This work has been supported by MH00423, MH42261, MH31154 and awards from the Scottish Rite Foundation.

References

Akbarian S, Bunney WE, Potkin SG, Wigal SB, Hagman JD, Sandman CA, Jones EG (1993) Altered distribution of nicotinamide-adenine dinucleotide phosphate-diaphorase cells in frontal lobe of schizophrenics implies disturbances of cortical development. Arch Gen Psychiatry 50:227–230

Angevine JB (1965) Time of neuron origin in the hippocampal region: an autoradiographic study in the mouse. Exp Neurol 13:1–70

Benes FM, Davidson J, Bird ED (1986) Quantitative cytoarchitectural studies of cerebral cortex of schizophrenics. Arch Gen Psychiatry 43:31–35

Benes FM, Turtle M, Farol P, SanGiovanni JP (1990) Myelination of cortico hippocampal relays continues through advanced adulthood in human brain. Arch Gen Psychiatry 51:477–484

Benes FM, McSparren J, Bird Ed, Vincent SL, SanGiovanni JP (1991a) Deficits in small interneurons in prefrontal and anterior cingulate cortex of schizophrenic and schizoaffective patients. Arch Gen Psychiatry 48:996–1001

Benes FM, Sorensen I, Bird ED (1991b) Morphometric analyses of the hippocampal formation in schizophrenic brain. Schizophr Bull 17:597–608

Benes FM, Sorensen I, Vincent SL, Bird ED, Sathi M (1992a) Increased density of glutamate-immunoreactive vertical processes in superficial laminae in cingulate cortex of schizophrenic brain. Cerebral Cortex 2:502–512

Benes FM, Vincent SL, Alsterberg G, Bird ED, SanGiovanni JP (1992b) Increased GABA-A receptor binding in superficial layers of cingulate cortex in schizophrenics. J Neurosci 12:924–929

Bogerts B, Meertz E, Schonfieldt-Bausch R (1985) Basal ganglia and limbic system pathology in schizophrenia: a morphometric study of brain volume and shrinkage. Arch Gen Psychiatry 42:784–791

Brown R, Colter N, Corsellis JAN, Crow TJ, Frith CD, Jagoe R, Johnstone EC, Marsh L (1986) Post-mortem evidence for structural brain changes in schizophrenia. Differences in brain weight, temporal horn area and parahippocampal gyrus width as compared with affective disorder. Arch Gen Psychiatry 43:36–42

Falkai P, Bogerts B (1986) Cell loss in the hippocampus of schizophrenics. Eur Arch Psychiatry Neurol Sci 236:154–161

Falkai P, Bogerts B, Rozumek M (1988) Cell loss and volume reduction in the entorhinal cortex of schizophrenics. Biol Psychiatry 24:515–521

Feighner JP, Robins E, Guze SB (1972) Diagnostic criteria for use in psychiatric research. Arch Gen Psychiatry 26:57–63

Fowler RC (1978) Remitting schizophrenia as a variant of affective disorder. Schizophr Bull 4:68–77

Jakob H, Beckmann H (1986) Prenatal developmental disturbances in the limbic allocortex in schizophrenics. J Neural Transm 65:303–326

Jeste D, Lohr JB (1989) Hippocampal pathologic findings in schizophrenia. Arch Gen Psychiatry 46:1019–1024

Kendler KS, Gruenberg A, Tsuang MT (1985) Psychiatric illness in first-degree relatives of schizophrenic and surgical control patients. Arch Gen Psychiatry 42:770–779

Kovelman JA, Scheibel AB (1984) A neurohistological correlate of schizophrenia. Biol Psychiatry 19:1601–1621

Levitt JJ, Tsuang MT (1988) The heterogeneity of schizoaffective disorder: Implications for treatment. Am J Psychiatry 145:926–936

Schizophrenia: Disease Entity, Disease Entities, or Domains of Psychopathology?*

W. T. Carpenter Jr, R. W. Buchanan, and B. Kirkpatrick[1]

Introduction

Schizophrenic patients, regardless of diagnostic concept and criteria applied, show a marked variability in age and type of onset, premorbid or early morbid adjustment, signs and symptoms, treatment response, and course of illness. Extensive between-subject variability is also observed in risk factors (e.g., genetic, gestational and birth complications, and season of birth) and associated biological and psychological features. Although individual variability contributes to differences in clinical manifestations, the observed clinical heterogeneity also underscores a conceptual issue of profound importance. Is schizophrenia a single disease, multiple diseases with certain core manifestations im common, or a varying combination of multiple pathophysiological processes, each leading to a different aspect of the syndrome?

The clinical heterogeneity of the syphilitic insanities supports the plausibility of the single disease concept of schizophrenia, and interesting and heuristically important unifying theories of schizophrenic pathophysiology have been proposed (Stevens 1973; Weinberger 1987; Carlsson and Carlsson 1990). However, no single disease process has yet been identified. This fact, combined with the varied occurrence of genetic and environmental risk factors, the failure to identify pathognomonic neuropathological or neurophysiological markers, and the manifest clinical heterogeneity, suggests that schizophrenia is a pathophysiologically heterogeneous disorder.

Two general models have been proposed that could explain the putative pathophysiological heterogeneity of schizophrenia. The first model posits that schizophrenia is a syndrome consisting of multiple disease entities. Each disease entity has its own unique pathophysiology that leads to the general manifestations of schizophrenia. The syndrome of mental retardation is an example of a medical condition that demonstrates the plausibility of different disease entities leading to the same core mental abnormality. It is also clear, in the field of psychiatry, that quite distinct etiologies can lead to similar psychotic manifestations. This model of heterogeneity has historic roots; Kraepelin (1971) and

* Supported in part by NIMH grant MH40279 and the National Alliance for Research in Schizophrenia and Depression (NARSAD).
[1] Maryland Psychiatric Research Center, Department of Psychiatry, University of Maryland School of Medicine, Baltimore, MD 21226, USA

Bleuler (1950) joined the putative disease entities of dementia praecox, catatonia, hebephrenia, paranoia, and schizophrenia under the rubric of dementia praecox or the "group of schizophrenias." However, in contrast to positing pathophysiological heterogeneity, both pioneers emphasized the pathophysiological features that the entities had in common, and Bleuler provided the still influential theory that a single pathological process (the dissociation within thought and between thought, affect, and behavior) was causative of the varied manifestations of schizophrenia. The study of schizophrenia in the disease entities or subtypes framework has been facilitated by the multiple criteria version of Crow's (1985) type-I and type-II schizophrenia and Andreasen's (1982) positive vs. negative schizophrenia.

The second pathophysiological heterogeneity model proposes that schizophrenia is comprised of discrete sign and symptom complexes, each the manifestation of an independent pathophysiological process. This approach emerged from work at the Washington Field Center of the WHO International Pilot Study of Schizophrenia (1974). The proposition that core symptoms defined nuclear schizophrenia, with a substantial reduction in heterogeneity, was disproved. On the contrary, it was observed that key components of the illness had only modest within-subject correlations, and that these correlations became negligible in the course of evaluation over 2 and 5 years. The status of any core feature tended to predict its own course, but only social dysfunction was predictive across features such as psychosis, duration of hospital care, occupation, and social functioning. This led to the 1974 model (Strauss et al. 1974) of three sign and symptom components of schizophrenia that merited independent study. For descriptive (rather than theoretical) purposes, the terms "positive symptoms", "negative symptoms", and "disorders observed in interpersonal relations" were introduced to name the three components (see Berrios 1985 for a review of the concept of negative symptoms).

The three-component hypothesis has now been demonstrated in most of the 15 relevant studies. Much of the following material is reviewed in detail elsewhere (Buchanan and Carpenter 1994). However, beginning with the report of Bilder et al. (1985), an important modification was required. Hallucinations and delusions were robustly associated with each other, but not with formal thought disorder. For this reason, we proposed five domains of interest: three represented a reordering (positive symptoms defined by hallucinations and delusions; negative symptoms defined by enduring and primary restricted affect, poverty of speech, diminished curiosity, sense of purpose and social drive; and disturbances in cognition/attention). Two additional domains were proposed, as their relation to the "big three" was unclear: incongruity of affect and behavior, and clinical neurological signs. Subsequent studies suggest that incongruity of affect is associated with cognitive impairment (Liddle 1987a; Liddle and Barnes 1990; Liddle et al. 1992; Mortimer et al. 1990; Brown and White 1992; Peralta et al. 1992) and aspects of neurologic manifestations are associated with the deficit or negative symptom domain (Liddle 1987a,b; Buchanan et al. 1990).

In the context of this second model of pathophysiological heterogeneity, the three primary domains of psychopathology (i.e., positive, negative, cognitive) are regarded as distinct at the level of etiopathophysiology, as well as at the level of disease manifestation. As an heuristic measure, each is treated as a disease and the schizophrenia syndrome comprises individuals with various combinations of these domains. Here it differs from the disease entities model of heterogeneity in an important conceptual and practical manner. Multiple disease processes are conceptualized as leading to discrete domains of psychopathology in the latter model, while in the former model distinct disease processes lead to similar core features. At the practical level, the disease entity model usually involves multiple identifying criteria between subtypes, and differences between putative disease entities might be caused by any or all of the defining criteria. In the domains model, a single domain defines the patient subgroup, and study designs can isolate this factor in within-schizophrenia studies that allow for more precise data interpretation.

The Crow typology can illustrate this distinction between the two models of pathophysiological heterogeneity. As noted above, multiple defining criteria can define two putative disease entities with certain common features (e.g., psychotic symptoms). Between-group differences may be attributable to differences in brain structure, treatment response, IQ, irreversible negative symptoms, etc. However, applied in the domains model context, schizophrenia with irreversible negative symptoms can be contrasted with schizophrenia without irreversible negative symptoms and differences interpreted as due to negative symptom pathology. Crow now favors this latter application (T. J. Crow 1991, personal communication), and would view criteria such as drug response and structural brain abnormalities as dependent variables used in hypothesis testing.

A final introductory remark concerns the methodological advantages associated with the assumption of pathophysiological heterogeneity (Carpenter et al. 1994). First, the assumption of heterogeneity enables within schizophrenia comparisons, the use of which provides a powerful approach to controlling key sources of artifact such as neuroleptic exposure and symptom duration and intensity. Second, the use of homogeneous subgroups strengthens schizophrenia/control comparisons by reducing type-II errors that result from study cohorts diluted with subjects not representing the subgroup. The assumption of heterogeneity, if incorrect, does not undermine schizophrenia comparisons with nonschizophrenic controls, since, under these circumstances, any subgroup would have the common pathophysiological process.

Dimensions, Domains, and Diseases

Heterogeneity reduction, analogous to the domains model, can proceed at several levels of the functioning human organism. For example, heterogeneity might be reduced by contrasting schizophrenia with and without eye-tracking abnormalities. The domains construct was introduced (Carpenter and Buchanan 1989) for the explicit purpose of facilitating heterogeneity reduction at the level of clinical symptoms. We chose the term "domain" since it was little used in psychopathology discussions and would not be as readily misunderstood as terms with histories of past usage. For example, "dimensions" often implies continuity with normalcy, and we wanted a term that implied a point of discontinuity representing the difference between disease and health. "Symptom clusters" might imply a relationship to cluster-analytic procedures; "components" or "factors" have the same disadvantage. "Subtypes" is not acceptable since it has many implications, and is usually applied in the context of model two – several disease entities leading to common syndrome manifestations.

A domain of psychopathology is a conceptually and empirically validated group of similar signs and/or symptoms of an illness that can be distinguished from other groups of signs and/or symptoms. Each domain is hypothesized to be the manifestation of a distinct pathophysiological process. A domain could itself have syndrome status with several pathophysiologies. This is clearly the case if not restricting subjects within the schizophrenia syndrome. For example, there are many causes of psychosis, but the psychosis of schizophrenia is more homogeneous. The domains approach emphasizes the investigation of discrete symptom complexes as compared with schizophrenic syndromes or subtypes. There is a paradigmatic shift away from the pathophysiology of schizophrenia to the investigation of the pathophysiology of the specific domain.

Validity of the Domains of Psychopathology in Schizophrenia

The validity of separating schizophrenic symptomatology into discrete symptom complexes will be considered from three vantages: empirical studies of schizophrenic patients designed to derive the number and constitution of psychopathological components; neurocognitive data as independent validation; and imaging data as independent validation. At the outset, we note that most of the relevant literature is confounded by two or three of the following: (a) a failure to restrict negative symptoms to those enduring features that are primary to schizophrenia, hence consistently weakening studies of negative symptom hypotheses; (b) the use of assessment procedures that imply a pattern of relationships between symptoms, thereby risking a halo effect and overestimating the independence of the three domains; and (c) the focusing of validation studies on positive and negative symptoms, with less clarity regarding cognitive/attentional dysfunction.

We have reviewed 15 articles that have examined symptom structure in schizophrenic patients. Eleven studies reported three-factor or -dimension solutions (Bilder et al. 1985; Kulhara et al. 1986; Liddle 1987a; Liddle and Barnes 1990; Liddle et al. 1992; Moscarelli et al. 1987; Mortimer et al. 1990; Gur et al. 1991; Arndt et al. 1991; Brown and White 1992; Minas et al. 1992). Three studies reported 4-factor solutions (Andreasen et al. 1992; Jorgensen and Parnas 1990; Peralta et al. 1992), and one study reported a seven-factor solution (Kay and Sevy 1990).

Most studies reported the presence of three factors: hallucinations and delusions, negative symptoms, and cognitive impairment. This conclusion is in line with the review by Klimidis et al. (1993).

There was greater variability of the factor on which the attentional impairment item loaded and the items that comprised the cognitive impairment factor. Bilder et al. (1985) reported that attentional impairment, along with positive formal thought disorder, bizarre behavior, and alogia, loaded on the cognitive impairment factor. Attentional impairment was associated with positive formal thought disorder in three other studies (Gur et al. 1991; Brown and White 1992; Peralta et al. 1992), but with negative symptoms in six studies (Andreasen and Olson 1982; Kulhara et al. 1986; Moscarelli et al. 1987; Jorgensen and Parnas 1990; Kay and Sevy 1990; Arndt et al. 1991). Similarly, in six studies, inappropriate affect, when part of the factor solution, was observed to load on the cognitive impairment factor (Liddle 1987a; Liddle and Barnes 1990; Liddle et al. 1992; Mortimer et al. 1990; Brown and White 1992; and Peralta et al. 1992).

Neurocognitive Assessment of the Domains

The work on neurocognitive assessments of schizophrenic domains of psychopathology has been in the context of correlations within schizophrenic cohorts. Three investigators have examined the validity of their factor solutions by examining the clinical correlates of the derived factors. Bilder et al. (1985) and Liddle (Liddle 1987b; Liddle and Morris 1991) used neurobehavioral measures to demonstrate the validity of their three-factor solutions. Bilder and colleagues reported that the three factors exhibited different patterns of correlations with neuropsychological tests of perceptual, language, memory, attentional, and executive functions. Liddle and colleagues have demonstrated, in three independent samples, a three-factor structure, with the symptoms segregating into three syndromes: reality distortion, disorganization, and psychomotor poverty (Liddle 1987a; Liddle and Barnes 1990; Liddle et al. 1992). In the initial sample, the disorganization syndrome score was negatively correlated with performance on tests of orientation, concentration, immediate recall, and word learning, while the psychomotor poverty syndrome score was negatively correlated with performance on tests of long-term memory, object naming, and conceptual thinking (Liddle 1987b). Both syndromes were significantly correlated with impairment on cortical neurological signs. In contrast, the reality distortion score did not correlate with any of the cognitive or neurological measures. In the second sample, they found that the psychomotor poverty and disorganization syndromes were related to impaired performance on frontal lobe tasks; psychomotor poverty was associated with slowness of mental activity and disorganization with impaired ability to inhibit established responses (Liddle and Morris 1991).

Our early work in this area focused on the contrast between schizophrenia with and without the deficit, with the deficit group having poorer performance on a range of neurocognitive tasks (Wagman et al. 1987). That these differences did not simply represent worse performance in a more severely ill group was suggested in a later study that found deficit patients characterized by impairments in sensory integration, with nondeficit schizophrenic patients beging similar to normal controls. Other neurological manifestations differentiated both groups from normals, but not from each other (Buchanan et al. 1990). In current studies (Buchanan et al., unpublished data), an interesting pattern of neurocognitive performance distinguishes deficit from nondeficit schizophrenia. We hypothesized that tasks emphasizing frontal and parietal functions would distinguish deficit schizophrenia, while temporal lobe-oriented tasks would be similar in the two schizophrenia groups and different from normal controls. The deficit group was significantly more impaired on selected frontal (Stroop Color-word Interference Test and Trails B) and parietal (Mooney Faces Closure) tasks, suggesting that deficit patients exhibit specific, and not generalized, impairments in frontal and parietal lobe functions. As hypothesized, there were no differences between deficit and nondeficit patients on tests emphasizing temporal lobe functions.

Taken together, these findings support the proposition of distinctive domains of psychopathology. They further suggest that the neural substrates involved in the symptom patterns are distinguishable.

Neural Basis of the Domains of Psychopathology

If the domains of psychopathology are a valid approach for classifying the subjective experience and psychological performance of schizophrenic patients, then the concept may be applied at the next level of inquiry – the neural substrate of schizophrenic symptomatology. In this framework, the investigation of the pathophysiology of schizophrenia becomes the separate investigation of the pathophysiology of hallucinations and delusions, negative symptoms, and cognitive impairment. Selected functional and structural imaging studies that have examined the neuroanatomy of hallucinations, delusions, formal thought disorder, and negative symptoms are presented.

Hallucinations and Delusions

The results of imaging, neurobehavioral, and postmortem studies examining schizophrenic patient/normal control differences have fairly consistently implicated temporal lobe structures (Suddath et al. 1989; Bogerts et al. 1985, 1990; Tamminga et al. 1992; Breier et al. 1992; Gruzelier et al. 1988; Saykin et al. 1991; Arnold et al. 1991). These studies provide indirect evidence for the involvement of temporal lobe structures in the production of hallucinations and delusions, as findings that generalize across study cohorts are most likely associated with those features; i.e., the psychotic domain is almost ubiquitous in schizophrenia. In contrast, although there are not yet good epidemiological data, formal thought disorder may occur in only about half of patients with schizophrenia, and primary, enduring negative symptoms in fewer than half (author's surmise).

Several positron emission tomography (PET) studies have examined the relationship between altered brain metabolism and hallucinations and delusions. Buchsbaum et al. (1982) found that schizophrenic patients with auditory hallucinations, as compared with normal controls, had higher glucose metabolic ratios in the tomographic slice that included Broca's area and temporal lobe auditory areas. However, there were no significant differences in regional brain metabolism between hallucinating and nonhallucinating patients. Volkow et al. studied patterns of glucose metabolism in 18 chronic schizophrenic patients (Volkow et al. 1987). Positive symptoms correlated with glucose metabolism in all of the cortical brain regions assessed, and hallucinations specifically correlated with increased right temporal lobe metabolism. There were no significant correlations between BPRS measures of conceptual disorganization and unusual thought content with any of the regional glucose metabolism measures. Tamminga et al. reported schizophrenic/normal control glucose metabolism differences in the hippocampus and anterior cingulate cortex (Tamminga et al. 1992). Although the relationship between hallucinations and delusions and glucose metabolism was not explicitly evaluated, the fact that deficit and nondeficit patients with similar levels of hallucinations and delusions exhibited similar hypometabolism in these limbic system structures supports the proposition that the altered glucose metabolism in these areas was associated with these symptoms. This interpretation is reinforced by the finding of decreased glucose metabolism in other brain regions associated with deficit symptomatology (see below).

In a series of studies, Cleghorn et al. examined glucose metabolism in schizophrenic patients who were actively hallucinating while undergoing PET scanning (Cleghorn et al. 1990, 1992). In the initial study, they found that the level of hallucinatory behavior was significantly correlated with relative metabolism in the anterior cingulate region, and that the pattern of positive correlations between right and left Broca's area, anterior cingulate region, and left superior temporal cortex differentiated hallucinating from nonhallucinating medicated schizophrenic patients (Cleghorn et al. 1990). There were no significant differences between hallucinating and nonhallucinating patients as regards the relative metabolism of any single brain region involved in speech and language. In a second study (Cleghorn et al. 1992), unmedicated hallucinating patients had lower glucose metabolism in the right and left posterior superior temporal gyrus and showed a trend towards higher glucose metabolism in right Broca's area. Hallucination scores were positively correlated with striatal and anterior cingulate metabolism. In contrast to the first study, the two groups were not differentiated by the patterns of correlations between different brain regions.

In a study examining the relationship between temporal lobe structure volumes and positive symptoms, auditory hallucinations were shown to be inversely correlated with anterior superior temporal gyral volume (Barta et al. 1990). There were no relationships observed between delusions or formal thought disorder and any of the temporal lobe structure volumes.

The results of the studies that have directly examined the relationship between hallucinations and brain region function or structure support the involvement of Broca's area, the superior temporal gyrus, hippocampus, and anterior cingulate gyrus in the production of hallucinations. The implication of these brain regions in the neuroanatomy of hallucinations is compatible with hyperdopaminergic mesolimbic theories of schizophrenic pathophysiology, but it is restricted to hallucinations and does not specify neurotransmitter involvement. The data are less clear with respect to the neuroanatomical localization of delusions. Although the studies reviewed in the previous section support the co-occurrence of hallucinations and delusions, the two imaging studies that examined delusions did not observe an association with temporal lobe metabolism or structure (Volkow et al. 1987; Barta et al. 1990).

Cognitive Impairment

There have been relatively few imaging studies that have examined the neuroanatomical localization of formal thought disorder and attentional/information processing impairments. Shenton et al. (1992) recently observed that the degree of thought disorder was inversely related to the volume of the left posterior superior temporal gyrus. These results – combined with the results of Barta et al. (1990), who found an association between hallucinations, but not formal thought disorder, with anterior superior temporal gyral volume – suggest that the neuroanatomical substrates for thought disorder and hallucinations may involve different aspects of the superior temporal gyrus.

The investigation of the neuroanatomical localization of attentional impairments has been hampered, in part, by the fact that attention is not a single function localized in a single brain region (Posner and Peterson 1990; Mirsky et al. 1991). Although there have been multiple studies documenting attentional impairments in schizophrenia (Strauss 1989; Neuchterlein 1991), few studies have attempted to delineate the brain regions involved in these impairments (Posner et al. 1988; Buchsbaum et al. 1990).

Deficit Symptoms

Lesions of the frontal lobes, especially the dorsolateral prefrontal cortex, or of the caudate result in apathy, abulia, emotional blunting, and decreased spontaneity of speech. These observations in nonschizophrenic subjects have led investigators to speculate on the involvement of these brain regions in the production of negative or deficit symptoms (Levin 1984; Carpenter et al. 1988;

Weinberger 1987). Several PET studies have documented a relationship between impaired frontal lobe performance and negative symptoms, and found negative symptoms to be related to reduced frontal lobe metabolism and rCBF (Volkow et al. 1987; Andreasen et al. 1992; Wolkin et al. 1992; Liddle et al. 1992).

The majority of studies examining the neuroanatomy of negative symptoms have not differentiated primary and enduring negative or deficit symptoms from secondary negative symptoms. We used PET to examine patterns of glucose metabolism in deficit and nondeficit patients. Deficit patients were characterized by significantly decreased glucose metabolism in the frontal and parietal cortices and thalamus (Tamminga et al. 1992), areas where the nondeficit chronic schizophrenic patients were similar to normal controls. These PET study results are consistent with previous neurobehavioral studies of deficit symptoms that demonstrated that deficit patients: (a) are more neurologically impaired, especially with respect to items sensitive to parietal lobe function (Buchanan et al. 1990); and (b) exhibit increased volitional saccadic latency, an eye-tracking measure sensitive to either impaired frontal or parietal eye-field function (Thaker et al. 1989). The implication of the frontal lobe in the production of deficit symptoms is in accord with previous studies of negative symptoms (Volkow et al. 1987; Andreasen et al. 1992; Wolkin et al. 1992; Liddle et al. 1992). The possible role of the parietal cortex and thalamus in the production of deficit symptoms has not been previously considered, although Liddle et al. (1992) have reported negative symptoms to be associated with decreased rCBF in the parietal association cortex. Finally, Liddle has provided further support for the independence of the three syndromes in a study simultaneously revealing a different pattern of cerebral blood flow associated with each symptom complex (Liddle et al. 1992).

CT and MRI studies of negative symptoms have been less successful in implicating specific brain regions in the production of negative symptoms. The majority of studies have failed to find a relationship between negative symptoms and structural changes in the prefrontal cortex (Pfefferbaum et al. 1988; Andreasen et al. 1986, 1990). In an MRI study examining prefrontal cortex and caudate volumes in deficit and nondeficit patients, prefrontal volume was bilaterally decreased in nondeficit patients; there were no significant differences in prefrontal volume between deficit patients and normal controls (Buchanan et al. 1993). The difference between deficit and nondeficit patients was due primarily to a reduction in white matter volume in the nondeficit patients. The failure to find an association between deficit symptoms and prefrontal structural changes is consistent with previous structural imaging studies of negative symptoms (Pfefferbaum et al. 1988; Andreasen et al. 1986, 1990). In our study, the caudate was bilaterally enlarged in deficit patients, with a trend for the right caudate to be differentially enlarged in deficit patients as compared with nondeficit patients and normal controls. The results of this study support the possibility of *structural* alterations in the caudate, but not prefrontal cortex, underlying deficit symptomatology. However, the prefrontal cortex may have a *functional* alteration, as suggested by the other data.

Future Directions

There is compelling evidence for the existence of three domains of psychopathology within schizophrenia. Rather than more studies describing and confirming this pattern, the field needs research designs which implement these findings and test specific pathophysiological hypotheses.

The concept of domains extends beyond investigations of the pathophysiology of schizophrenia and can also be used in investigations of schizophrenic etiology and treatment. In family or genetic studies, the domains would be hypothesized to represent independent phenotypes and could be used in conjunction with syndromal diagnoses to characterize study samples. The use of domains in treatment trials would provide more detailed information on the range of the therapeutic efficacy of pharmacological agents (Carpenter and Buchanan 1989). For example, although there is extensive evidence on the general therapeutic efficacy of antipsychotic agents for positive symptoms, there is relatively little evidence for the effect of these agents on primary negative symptoms. Improvement in secondary negative symptoms cannot be interpreted as demonstrating treatment efficacy for the deficit syndrome. Finally, animal models can be developed for each of the domains and, we would propose, will usually not be able to model more than one domain. These models can be used not only to investigate the pathophysiology of each domain, but also to aid in the development of specific pharmacological treatments for each domain.

References

Andreasen NC (1982) Negative symptoms in schizophrenia: definition and reliability. Arch Gen Psychiatry 39:784–788

Andreasen NC, Olsen S (1982) Negative v positive schizophrenia; definition and validation. Arch Gen Psychiatry 39:789–794

Andreasen N, Nasrallah HA, Dunn V et al (1986) Structural abnormalities in the frontal system in schizophrenia: a magnetic resonance imaging study. Arch Gen Psychiatry 43:136–144

Andreasen NC, Flaum M, Swayze VW (1990) Positive and negative symptoms in schizophrenia. Arch Gen Psychiatry 47:615–621

Andreasen NC, Rezai K, Alliger R et al (1992) Hypofrontality in neuroleptic-naive patients and in patients with chronic schizophrenia. Assessment with xenon 133 single-photon emission computed tomography and the Tower of London. Arch Gen Psychiatry 49:943–958

Arndt S, Alliger RJ, Andreasen NC (1991) The distinction of positive and negative symptoms; the failure of a two-dimensional model. Br J Psychiatry 158:317–322

Arnold SE, Hyman BT, Van Hoesen GW et al (1991) Some cytoarchitectural abnormalities of the entorhinal cortex in schizophrenia. Arch Gen Psychiatry 48:625–632

Barta PE, Pearlson GD, Powers RE et al (1990) Auditory hallucinations and smaller superior temporal gyral volume in schizophrenia. Am J Psychiatry 147:1457–1462

Berrios GE (1985) Positive and negative symptoms and Jackson. Arch Gen Psychiatry 42:95–97

Bilder RM, Mukherjee S, Rieder RO et al (1985) Symptomatic and neuropsychological components of defect states. Schizophr Bull 11:409–419

Bleuler E (1950) Dementia praecox or the group of schizophenias. (J Zinken, translator). International Universities Press, New York
Bogerts B, Meertz E, Schonfeldt-Bausch R (1985) Basal ganglia and limbic system pathology in schizophrenia: a morphometric study of brain volume and shrinkage. Arch Gen Psychiatry 42:784–791
Bogerts B, Ashtari M, Degreef G et al (1990) Reduced temporal limbic structure volumes on magnetic resonance images in first-episode schizophrenia. Psychiatry Res 35:1–13
Breier A, Buchanan RW, Elkashef A et al (1992) Brain morphology and schizophrenia: an MRI study of limbic, prefrontal cortex and caudate structures. Arch Gen Psychiatry 49:921–926
Brown KW, White T (1992) Syndromes of chronic schizophrenia and some clinical correlates. Br J Psychiatry 161:317–322
Buchanan RW, Carpenter WT (1994) Domains of psychopathology: an approach for the reduction of heterogeneity in schizophrenia. J Nerv Ment Dis (in press)
Buchanan RW, Kirkpatrick B, Heinrichs DW et al (1990) Clinical correlates of the deficit syndrome of schizophrenia. Am J Psychiatry 147:290–294
Buchanan RW, Breier A, Kirkpatrick B et al (1993) Structural abnormalities in deficit vs non-deficit schizophrenia. Am J Psychiatry 150:59–65
Buchsbaum MS, Ingvar DH, Kessler R et al (1982) Cerebral glucography with positron tomography; use in normal subjects and in patients with schizophrenia. Arch Gen Psychiatry 39:251–259
Buchsbaum MS, Neuchterlein KH, Haier RJ et al (1990) Glucose metabolic rate in normals and schizophrenics during the Continuous Performance Test assessed by positron emission tomography. Br J Psychiatry 156:216–227
Carlsson M, Carlsson A (1990) Interactions between glutamatergic and monoaminergic systems within the basal ganglia: implications for schizophrenia and Parkinson's disease. In: Swanson G (ed) Trends in neurosciences, reference edn, vol 13. Elsevier, Cambridge, pp 272–276
Carpenter WT, Buchanan RW (1989) Domains of psychopathology relevant to the study of etiology and teratment in schizophrenia. In: Schulz SC, Tamminga CA (eds) Schizophrenia: scientific progress. Oxford Press, New York, pp 13–22
Carpenter WT, Heinrichs DW, Alphs LD (1988) Deficit and nondeficit forms of schizophrenia: the concept. Am J Psychiatry 145:578–583
Carpenter WT, Buchanan RW, Kirkpatrick B et al (1994) Strong inference, theory falsification, and the neuroanatomy of schizophrenia. Arch Gen Psychiatry (in press)
Cleghorn JM, Garnett ES, Nahmias C et al (1990) Regional brain metabolism during auditory hallucinations in chronic schizophrenia. Br J Psychiatry 157:562–570
Cleghorn JM, Franco S, Szechtman B et al (1992) Toward a brain map of auditory hallucinations. Am J Psychiatry 149:1062–1069
Crow TJ (1985) The two-syndrome concept: origins and current status. Schizophr Bull 11:471–486
Gruzelier J, Seymour K, Wilson L et al (1988) Impairments on neuropsychologic tests of temporohippocampal and frontohippocampal functions and word fluency in remitting schizophrenia and affective disorders. Arch Gen Psychiatry 45:623–629
Gur RE, Mozley PD, Resnick SM et al (1991) Relations among clinical scales in schizophrenia. Am J Psychiatry 148:472–478
Jorgensen A, Parnas J (1990) The Copenhagen high-risk study; premorbid and clinical dimensions of maternal schizophrenia. J Nerv Ment Dis 178:370–376
Kay SR, Sevy S (1990) Pyramidical model of schizophrenia. Schizophr Bull 16:537–544
Klimidis S, Stuart GW, Minas IH et al (1993) Positive and negative symptoms in the psychoses. Re-analysis of published SAPS and SANS global ratings. Schizophr Res 8:11–18
Kraepelin E (1971) Dementia praecox and paraphrenia. (Translated from the original, 1919, by RM Barclay). Krieger, New York

Kulhara P, Kota SK, Joseph S (1986) Positive and negative subtypes of schizophrenia; a study from India. Acta Psychiatr Scand 74:353–359
Levin S (1984) Frontal lobe dysfunctions in schizophrenia. II. Impairments of psychological and brain functions. J Psychiatr Res 18:57–72
Liddle PF (1987a) The symptoms of chronic schizophrenia; a reexamination of the positive-negative dichotomy. Br J Psychiatry 151:145–151
Liddle PF (1987b) Schizophrenic syndromes, cognitive performance and neurological dysfunction. Psychol Med 17:49–57
Liddle PF, Barnes TRF (1990) Syndromes of chronic schizophrenia. Br J Psychiatry 157:558–561
Liddle PF, Morris DL (1991) Schizophrenic syndromes and frontal lobe performance. Br J Psychiatry 158:340–345
Liddle PF, Friston KJ, Frith CD et al (1992) Patterns of cerebral blood flow in schizophrenia. Br J Psychiatry 160:179–186
Minas IH, Stuart GW, Klimidis S et al (1992) Positive and negative symptoms in the psychoses: multidimensional scaling of SAPS and SANS items. Schizophr Res 8:143–156
Mirsky AF, Anthony BJ, Duncan CC et al (1991) Analysis of the elements of attention: a neuropsychological approach. Neuropsychology Rev 2:109–145
Mortimer AM, Lund CE, McKenna PJ (1990) The positive: negative dichotomy in schizophrenia. Br J Psychiatry 157:41–49
Moscarelli M, Maffei C, Cesana BM et al (1987) An international perspective on assessment of negative and positive symptoms in schizophrenia. Am J Psychiatry 144:1595–1598
Neuchterlein KH (1991) Vigilance in schizophrenia and related disorders. In: Steinhauer SR, Gruzelier JH, Zubin J (eds) Handbook of schizophrenia, 5. Elsevier, Amsterdam, pp 397–443
Peralta V, de Leon J, Cuesta MJ (1992) Are there more than two syndromes in schizophrenia? A critique of the positive-negative dichotomy. Br J Psychiatry 161: 335–343
Pfefferbaum A, Zipursky RB, Lim KO et al (1988) Computed tomographic evidence for generalized sulcal and ventricular enlargement in schizophrenia. Arch Gen Psychiatry 45:633–640
Posner MI, Peterson SE (1990) The attention system of the human brain. Annu Rev Neurosci 13:255–42
Posner MI, Early T, Reiman E et al (1988) Asymmetries in hemispheric control of attention in schizophrenia. Arch Gen Psychiatry 45:814–821
Saykin AJ, Gur RC, Gur RE et al (1991) Neuropsychological function in schizophrenia: selective impairment in memory and learning. Arch Gen Psychiatry 48:618–624
Shenton ME, Kikinis R, Jolesz FA et al (1992) Abnormalities of the left temporal lobe and thought disorder in schizophrenia. N Engl J Med 327:604–612
Stevens JR (1973) Neuroanatomy of schizophrenia. Arch Gen Psychiatry 29:177–189
Strauss JS, Carpenter WT, Bartko JJ (1974) The diagnosis and understanding of schizophrenia. III. Speculations on the processes that underlie schizophrenic symptoms and signs. Schizophr Bull 11:61–75
Strauss ME (1989) Cognitive markers for schizophrenia: an appraisal. In: Schultz SC, Tamminga CA (eds) Schizophrenia: scientific progress. Oxford Press, New York, pp 103–109
Suddath RL, Casanova MF, Goldberg TE et al (1989) Temporal lobe pathology in schizophrenia: a quantitative magnetic resonance imaging study. Am J Psychiatry 146:464–472
Tamminga CA, Thaker GK, Buchanan RW et al (1992) Limbic system abnormalities identified in schizophrenia using PET/FDG, and neocortical alterations with the deficit syndrome. Arch Gen Psychiatry 49:522–530

Thaker G, Kirkpatrick B, Buchanan RW et al (1989) Oculomotor abnormalities and their clinical correlates in schizophrenia. Psychopharmacol Bull 25:491–498

Volkow ND, Wolf AP, Van Gelder P et al (1987) Phenomenological correlates of metabolic activity in 18 patients with chronic schizophrenia. Am J Psychiatry 144: 151–158

Wagman AMI, Heinrichs DW, Carpenter WT (1987) The deficit and nondeficit forms of schizophrenia: neuropsychological evaluation. Psychiatry Res 22:310–330

Weinberger DR (1987) Implications of normal brain development for the pathogenesis of schizophrenia. Arch Gen Psychiatry 44:660–669

Wolking A, Sanfilipo M, Wolf AP (1992) Negative symptoms and hypofrontality in chronic schizophrenia. Arch Gen Psychiatry 49:959–965

World Health Organization (1974) Report of the International Pilot Study of Schizophrenia, vol 1. WHO Press, Geneva

Psychotic Continuum or Disease Entities? The Critical Impact of Nosology on the Problem of Aetiology

T. J. Crow

With the privilege of making the first and the last contributions of the symposium I will attempt to relate the end to the beginning – to draw from our discussions of nosological problems some lessons for the central issue of aetiology. As my text I take the principle of the thirteenth century Franciscan William of Ockham, who is described as a "nominalist" – that is to say as someone who believes that the labels that we use are arbitrary and do not necessarily represent underlying realities. Ockham's principle (the razor) is variously stated as "Entities are not to be multiplied beyond necessity" or "It is vain to do with more what can be done with less".

Some such principle is surely fundamental to scientific progress. What we must strive for is the most economical description of what we observe, in the hope that this will lead us to an understanding of the underlying natural phenomena. Such considerations are relevant to the question of heterogeneity vs. continuity, especially as it relates to aetiology. All will agree that what we see in the clinical picture of psychosis is great variation. The question we have to address is whether this diversity is intrinsic to aetiology. Are the mechanisms of causation those that are generating diversity in the general population as well as in disease, or are there multiple and unrelated causes of psychosis?

Is Schizophrenia a Neurohumoural Disorder?

Daniel van Kammen (this volume), given the difficult remit of discussing biochemical heterogeneity, concluded that monoamine hypotheses have not produced biochemical subgroups or clarified aetiology and suggests that "this failure may be due to prematurely concluding that biochemical measures reflect aetiology rather than pathophysiology". The point is well taken but there may be a more fundamental question – should we expect schizophrenia to be a disorder of neurohumoural transmission? The concept of "functional psychosis" has made attractive the notion that there are diseases of which the primary disturbance is neurochemical rather than structural; the discovery of the neuroleptic drugs, and subsequently of dopamine receptor blockade as their

---

University Department of Psychiatry, Warneford Hospital, Oxford OX3 7JX, UK

apparent mechanism of action encouraged this view. Much research endeavour has been spent on possible disturbances of neurohumoural transmission in the hope that they will identify a fundamental defect, and a number of the contributions to this symposium (e.g., in relation to positron emission tomography (PET) and other studies of dopamine receptors) made use of the concept. But is it likely that a simple neurotransmitter dysfunction could explain the occurrence of psychosis? As Dan van Kammen rightly remarks, a monoamine disturbance (if one exists) could be steps removed from the primary cause. That many illnesses respond to neuroleptic medication is an important fact but it has not led us to an understanding of aetiology. Nor has the attempt to form subgroups of illness on the basis of biochemical indices contributed much.

Might dividing illness on the basis of age of onset, cognitive changes or negative syptoms have greater success in identifying aetiologically distinct subgroups? As Eaton et al. (1992) have shown, the relationship between age of onset and outcome is a lawful one, but there is no evidence that those with early onset have a different form of illness. Age of onset is one of the variables that must be taken into account in understanding the nature of the illness, but it is not a magic key to aetiological substratification.

What Can We Learn from Neuropathology?

Francine Benes addresses some of the same issues through histopathology: but it is a needle in the haystack problem. Whereas the number of neurotransmitters, and therefore the number of possible neurohumoral theories may be assumed to be finite, this is not true of indices of morphology. It is therefore puzzling in Benes' paper that so many differences between patients with schizophrenia (and with different types of illness) and other individuals are readily demonstrated. What is the relationship between these diverse findings and how are they related to a disease process? One is reminded of the sceptical contribution of Dunlap (1924) with respect to earlier claims of specific changes in the cerebral cortex. Questions that one needs to ask in this type of investigation are how reliable is the technique and how does one standardize it from one brain to another? How does one ensure that the sample of brain that one has from, say, cingulate cortex is comparable between brains, and how does one allow for differences in brain weight and morphology? Obviously, progress depends upon replicability. With respect to gross changes in the brain in schizophrenia, there is perhaps now a measure of agreement – for example, that the ventricles are enlarged (Daniel et al. 1991) and the parahippocampal gyrus is reduced in width (Bogerts et al. 1985; Brown et al. 1986; Altschuler 1990). The way ahead may be to focus on relatively simple indices of morphology until their relationship (if any) to the disease process is clearly established and then to ask, at a microscopic level, what the cellular correlates of these changes may be.

The Importance of Nosology

Laboratory investigations of heterogeneity, whether chemical or anatomical, are based upon the assumption that such heterogeneity exists and that clinical boundaries can be reliably defined. But this is just what is at issue. In different ways the contributions of Will Carpenter and Nancy Andreasen illuminate, and indicate how fundamental is, the problem. The contrast of these presentation is instructive. Carpenter is interested in heterogeneity within the schizophrenia spectrum and Andreasen with defining the borders (if any) between schizophrenia and other illnesses.

Will discusses the distinction between dimensions, domains and disease entities and favours domains as the most promising correlate of "aetiopathophysiology". But he also refers to Bleuler's "still influential theory that a single pathological process (the dissociation within thought and between thought, affect and behaviour) was causative of the varied manifestations of schizophrenia". And he later describes the psychotic domain (hallucinations and delusions) as "almost ubiquitous" in schizophrenia. To blur the distinction between pathophysiology and aetiology (as in the use of the term "aetiopathophysiology") is perhaps unfortunate here. To speculate about separate pathophysiological processes (on the basis of imaging studies, for example) is one thing, but to postulate separate aetiologies for different domains of pathology when we have no single aetiology that we can be confident of which is another. One assumes, but he does not make it entirely clear, that Will does not expect each "domain" to have a different aetiology.

Nancy is concerned with the definition of the schizophrenia spectrum itself and addresses particularly the boundary with affective disorder. She notes that by DSM-III criteria the presence of a full affective syndrome is used an an exclusionary criterion "but (as anyone who works closely with 'schizophrenic' patients knows only too well), some have in fact met criteria for a full affective syndrome sometime during the course of the illness, and some are given the diagnosis of affective disorder on their first or second admission before the course and symptoms become clearly characteristic of schizophrenia . . . the boundaries between these disorders are not as clear in real life as . . . in diagnostic manuals . . . Controversy remains as to the importance of a variety of defining aspects . . . These controversies are neither academic nor trivial". She goes on to say that "there is no truly certain way to define schizophrenia" and concludes that the schizophrenia spectrum "almost certainly includes schizoaffective disorder . . . (and) . . . should probably also include at least the psychotic forms of mania and depression, since mood incongruent psychotic mania and depression are both essentially identical to older concepts of schizophrenia". These are radical proposals but, if mood-incongruent, affective psychoses are to be allowed in, why should not mood-congruent psychoses (as Andreasen's Table 1 suggests) also be included?

Tables 4 and 5 of Andreasen's presentation extend these interesting thoughts. But, in passing, it is worth noting what are described in Table 4 as three dimensions are what Will Carpenter would describe as domains. The distinction between the two is not all that clear.

The relationship between dimensions and diagnoses, also discussed in detail by Marneros in his contribution to this volume, is at the heart of the matter. Which (dimensions or diagnoses) has greater reality? Which is to be accorded greater priority in attempts to understand aetiology?

Are There Any Disease Entities?

Perhaps we should grasp the nettle and conclude that the conditions we are concerned with are indeed continuous. There are no defining features such as would be necessary to establish the existence of discrete diagnostic entities. The alternative is that we have deviations which, when they become identifiable as pathological (i.e., disadvantageous), can be used as markers along a continuum or hierarchy of psychotic illness. An attempt at an outline of such a descriptive system is shown in Fig. 1.

According to this system the markers define regions of pathology of increasing severity; thus illnesses with nuclear symptoms are generally more severe than those which include mood-congruent delusions without nuclear symptoms, and illnesses with evidence of the defect state (primary negative symptoms) are more incapacitating than those without such symptoms. Two further points are worth noting: more severe illnesses, and their associated markers, generally occur at progressively younger ages; and the range of affective disturbance becomes progressively more restricted as one moves from the affective to the schizophrenic "domains" of the spectrum.

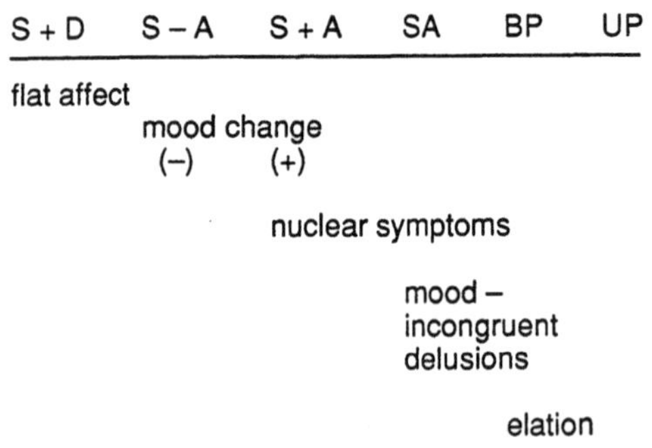

Fig. 1. A continuum concept with possible transitional markers. *UP*, unipolar illness; *BP*, bipolar illness; *SA*, schizoaffective illness; *S+A*, schizophrenic illness with significant mood disturbance; *S−A*, schizophrenic illness without significant mood disturbance; *S+D*, presence of affective flattening in the defect state

E. Bleuler's Legacy

The debate concerning continua (or a continuum) vs. disease entities deserves a historical context. The concept of schizophrenia owes much to E. Bleuler, and his formulations are referred to by both Will Carpenter and Nancy Andreasen. But the references they make to a "disease entity" concept that Bleuler is thought to have adhered to are revealingly different. Thus Andreasen states that "Bleuler (Bleuler 1950) broadened the concept substantially by introducing the term 'schizophrenia', emphasizing cross-sectional phenomenology is more important than outcome, and arguing that the disease should be defined in terms of 'fundamental' symptoms such as affective blunting or avolition rather than the positive symptoms of psychosis, which he viewed as 'accessory' and nonspecific." Carpenter (as already noted) asserts that "Bleuler provided the still influential theory that a single pathological process (the dissociation within thought and between thought, affect, and behaviour) was causative of the varied manifestations of the disease". Unless one believes that dissociation leads directly to affective blunting and avolition, these views are not consistent.

K. Schneider's Critique

It seems that Bleuler himself was confused. In a telling passage K. Schneider (1974) addressed Bleuler's claim to have identified a disease entity:

"Bleuler says: 'Certain symptoms of schizophrenia are present at all times and in every case'. Here he is referring to so-called 'basic symptoms'. 'Accessory' symptoms, on the other hand, may also occur in other illnesses. Bleuler's 'basic' symptoms are, however, a theoretical postulate: in order to be recognised with confidence, they must have attained a certain degree of severity', This means, however, that they are *not* present at all times and in every case', as Bleuler himself expressly concedes. It is not clear how these 'basic' symptoms are related to the 'primary' symptoms, which he mentions for the first time only in his theoretical discussion. Nor is it clear which phenomena he regards as essential 'basic' symptoms and which as primary. The contradictions here seem to me irreconcilable. The only basic symptoms which he lists initially are disturbed association, disturbed affect and autism. These symptoms are then subdivided into disturbances of simple and complex functioning. Among the disturbances of simple functioning he now includes, surprisingly, ambivalence: autism, on the other hand, is listed under complex functions, along with much else."

"One might expect Bleuler's primary symptoms ('the indispensable clinical features of an illness') to be more or less the same as the basic symptoms, but, again surprisingly, we find they are quite different. Of the basic symptoms, only disturbed association counts as primary and even then there are reservations: disturbed affect, ambivalence, autism are not primary but secondary symptoms, disturbed affect being listed as provisional. This and much more is inconsistent, illogical, and unsystematic. There is not even any clear indication as to which symptom should, even theoretically, be present at all times. Disturbed association seems to take precedence over all other basic and primary symptoms. As already pointed out, however, it clearly does not have to be present in every case of schizophrenia... Bleuler has deduced a great deal from his 'disturbed association', a vague and ambiguous concept which does little justice to the richness and power of schizophrenic symptomatology."

Schneider here has effectively demolished Bleuler's claim to have established a disease entity. Bleuler was not able to solve the problem for us, any more than the concept of an underlying disturbance of dopaminergic transmission did so at a later date. Schneider, adopting a leaner and more parsimonious approach, was looking for more dependable indices.

The Central Importance of the WHO Studies

I now want to argue that the consistencies of the natural phenomena of psychosis, and therefore the constraints on our theories, are greater than the foregoing history would lead us to believe. The World Health Organization Ten Country Study of Incidence (Jablensky et al. 1992) demonstrated that when restrictive criteria (based upon Schneider's first rank symptoms) are adopted, onsets of illness are similar in societies which differ widely in their climatic, social and industrial environments. The authors conclude that "schizophrenic illnesses are ubiquitous, appear with similar incidence in different cultures and have clinical features that are more remarkable by their similarity across cultures than their differences". Although with broader criteria there were apparent differences between centres in incidence (Fig. 2), it is plausible that such criteria identify illnesses of lesser degrees of severity which are therefore less certain (from variations in the availability of services, or traditions of care-seeking) to be enumerated. The particular contribution of Schneider may have been to identify a level of psychological disturbance at which it is highly likely that, if the subject experiences it for the first time, he will seek help.

But if schizophrenia is universal in this sense it surely differs from other common conditions such as ischaemic heart disease, hypertension and diabetes. These disorders vary significantly in different societies and such variations are held to reflect an environmental influence. In schizophrenia (at least when restrictively defined) we have no evidence for such an influence. Perhaps one should conclude that, in a sense that is not true of conditions such as ischaemic heart disease and diabetes, schizophrenia is a disorder of humanity. And if this conclusion can be drawn concerning the core syndrome perhaps it is true also of other major psychiatric conditions – schizoaffective and affective psychoses – with which schizophrenic disorders form a continuum.

The second major constraint on our theories is age of onset. Schizophrenia (along with the other psychoses) occurs throughout the reproductive phase of life and is associated with a fertility disadvantage. Schizophrenic patients have

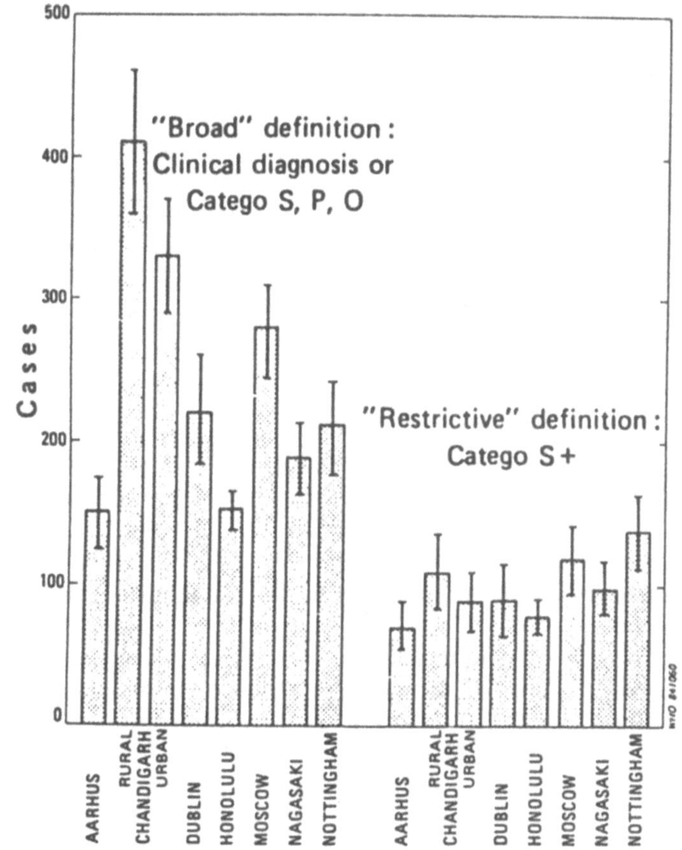

Fig. 2. The main findings of the WHO Ten Country Study of Incidence (Jablensky et al. 1992). *S*, schizophrenic psychosis; *P*, paranoid psychosis; *O*, borderline or doubtful psychosis

fewer children than the population mean, and this is particularly true of male patients. An evolutionary explanation of the persistence of the gene or genes therefore is required.

The Contribution of Morphology

Evidence for a degree of homogeneity in psychosis also comes from morphological studies. Ventricle-to-brain ratio is elevated in patients relative to normal and medical comparison groups, but within the patient group variance is not increased and there is no evidence of bimodality (Daniel et al. 1991). In some sense structural change is characteristic of the schizophrenic "disease process" in general, not of a subgroup. Other findings are that brain size is reduced (Brown et al. 1986; Bruton et al. 1990; Pakkenberg 1987) and there may be a reduction in cortical mass (Zipursky et al. 1992; Harvey et al. 1993). In addition, development of asymmetry is arrested (Crow et al. 1989a,b; Falkai et al. 1992). These strands of evidence can be drawn together in the proposal (Crow 1993) that the primary change is a genetically determined arrest of development of asymmetry and that this is secondarily associated with a failure of development of recently evolved association areas of cortex and an increase (following from cortical arrest) in ventricular space. According to this conclusion, schizophrenia (and by extension other psychotic illness) is intimately associated with the recent evolutionary increase in brain size.

Two Postulates

On this background, I propose two generalizations concerning the nature and origins of psychotic illness. These can be considered as working hypotheses or "postulates":

Postulate 1: there are no environmental causes.

Postulate 2: there are no disease entities, only deviations along dimensions of psychological variation.

These postulates have a certain resemblance to stages in evolutionary thinking. The first is along the lines of Weissman's insistence that there is only "hard" inheritance – no Lamarckian element of environmental influence should be conceded. According to this postulate discordance for illness in monozygotic twins is to be attributed to epigenetic or even genetic differences, not to the environment. The second postulate bears a resemblance to Darwin's insight that species are not fixed.

But if there are no true disease entities there is no basis for isolating one part of the pathological spectrum – the whole range must be considered. But in this case one must ask where does the variation come from? It must be genetic and some of it is disadvantageous. What evolutionary pressures are continuing to generate such diversity?

The Dimensions of Variation: The Sexual Dimorphism for Cerebral Asymmetry

Following the lead set by Kretschmer one can ask to which dimension of variation in the normal population does the diversity of psychosis (from schizophrenia to manic-depressive illness) relate? Figure 3 makes the proposal that the dimension, which also relates to age of onset and a sex difference, is one of "sociability – emotionality".

The range of social competence and the capacity for emotional expression develops with age. Part of what we see in psychosis are deviations in this capacity, restricted at the schizophrenic end and exaggerated at the affective end, but there are clearly other components. Where does the sex difference arise from? It must surely have something to do with the sexual dimorphism for

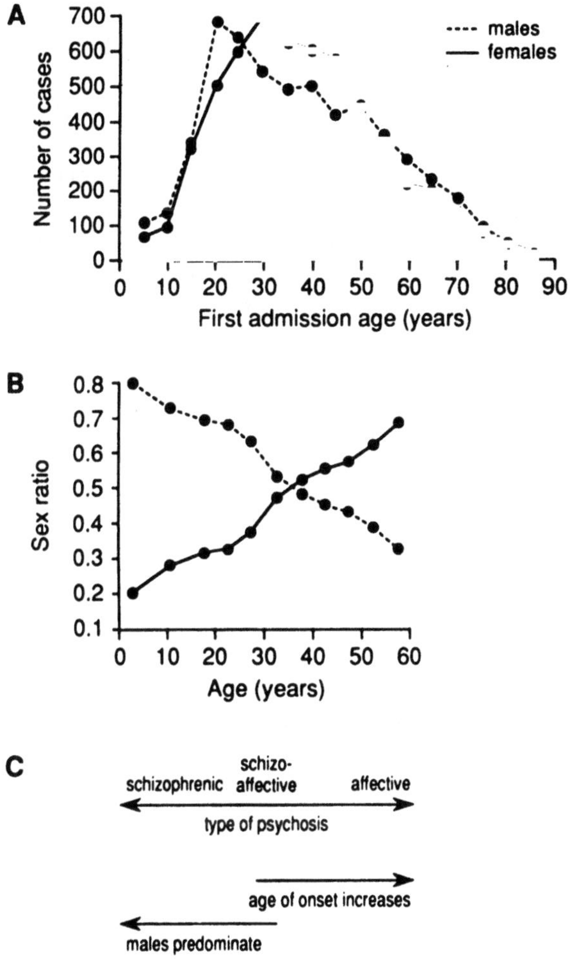

Fig. 3A–C. Relation between age, sex, and onset of psychosis. **A** Age at first hospital admission according to L. S. Penrose (*dashed line*, males; *continuous line*, females). **B** Sex ratio for schizophrenic type psychoses at different ages adapted from D. Rosenthal (*dashes*, % male; *continuous line*, % female). **C** The continuum notion of psychosis, illustrating the relation between form of illness, age of onset, and range of sociability/emotionality in the normal population (Crow 1993)

cerebral asymmetry – the likely correlate of the difference in means between the sexes for cognitive function – females having greater verbal fluency and males better spatial ability.

Sexual Selection

Such a difference must arise by the mechanism of sexual selection (Darwin 1871). Buss et al. (1990) have shown a cross-culturally robust difference between the sexes in criteria for mate choice. Whereas both sexes rate intelligence, and kindness and understanding highly, males rate physical attractiveness higher than women and women rate earning capacity higher than men. These differences can be understood in terms of the differing interests of the sexes in procreation – females are interested in good genes and paternal investment – for this established success may be the most appropriate indicator; males are interested in fecundity, and for this physical attractiveness may be the best index of the relevant attributes of health and youth. If relative youth and established success indeed influence mating decisions, a mean difference between the sexes would be predicted in age at procreation. This is borne out by data from the UN Demographic Year Book – females are a mean 3 to 4 years younger than males at the time of marriage (Crow 1993).

Three sex differences – criteria for mate choice, age at marriage, and age of onset of psychosis – thus are stable across cultures and may be related. Perhaps the difference between the sexes in criteria for choosing a mate, acting upon a genetic innovation (the potential for a difference between the hemispheres in rate of development) in homo sapiens, has generated a dimension of variation in cognitive personality structure. Different variants have differing survival values in different (social) environments and are subject to continuing (sexual) selection, the boundary conditions which we recognize as psychosis being associated with a particular disadvantage.

X-Y Homologous Genes and the Timing of Development

Timing is central. If the male brain has a greater mean asymmetry than the female, this must be determined by genes that act early in development. But the key to their selection may be the age at mate choice (Fig. 4).

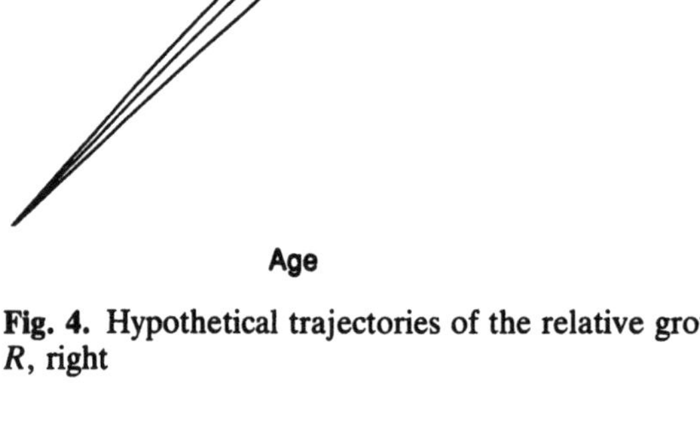

Fig. 4. Hypothetical trajectories of the relative growth of the two hemispheres. *L*, left; *R*, right

Where is the gene(s) that determines asymmetry (variants of which according to this theory predispose to psychosis)? Two lines of evidence support a location on the X and Y chromosomes:

1. The neuropsychological profile of individuals with sex chromosome aneuploidies (Crow 1993)
2. A tendency for handedness within sibling pairs to be linked to sex (M.C. Corballis et al., unpublished data)

A gene represented in homologous but differing form on both X and Y chromosomes could account for the sexual dimorphism in cerebral asymmetry and could be relevant to the sex difference in age of onset of psychosis.

Timing

Early onset of psychosis predicts higher probability of relapse, as also do male sex and single status (Eaton et al. 1992). But when age of onset is taken into account the predictive power of sex and marital status disappear. Thus, the phenomena of psychosis are linked to a mechanism that times an aspect of CNS development related to, but not dependent upon, sex. The relative development of the two hemispheres (Fig. 4) could reflect the same process.

One can envisage that a large number of trajectories of development are possible with a mean difference, but considerable overlap in range, in the two sexes. Perhaps there is a trade-off between brain size and degree of asymmetry, with the genetic factor influencing the point at which growth reaches a plateau. The difference between the sexes might then be represented by a quantitative distribution (coded perhaps by a number of a repeated sequence in a regulatory element) on the X and Y chromosomes, respectively. Such genetic variation would be subject to sexual selection acting differentially on the sex chromosomes.

Conclusions

Two working hypotheses concerning the origins of psychosis are proposed:

Postulate 1: an environmental contribution is absent.

Postulate 2: there are no disease entities but the psychoses can be regarded as "boundary" conditions of continuous variation that is present in the general population.

Acceptance of these postulates leads on to the conclusion that psychotic disorder arises, as Kretschmer and later Kraepelin suggested, as a consequence of genetic variation that is specific to, and related to the recent evolution of, homo sapiens. The key genetic change is the potential for the cerebral hemispheres to develop with a degree of independence. Sexual selection, acting on this substrate, has generated a dimension of variation with considerable overlap but a mean difference between the sexes.

Such variation can most parsimoniously be explained on the assumption that there is a single gene that exists in homologous but differing forms on both X and Y chromosomes and exerts an influence on the trajectory and/or timing of the plateau of hemispheric growth. Such a gene could account for the sexual dimorphism in cerebral asymmetry, and the sex differences in verbal fluency and spatial ability, and in age of onset of schizophrenia.

References

Altschuler LL, Casanova MF, Goldberg TE, Kleinman JE (1990) The hippocampus and para-hippocampus in schizophrenic, suicide and control brains. Arch Gen Psychiatry 47:1029–1034

Bleuler E (1950) Dementia praecox or the group of schizophrenias (translated by J. Zinkin). International University Press, New York

Bogerts B, Meertz E, Schonfeldt-Bausch R (1985) Basal ganglia and limbic system pathology in schizophrenia: a morphometric study of brain volume and shrinkage. Arch Gen Psychiatry 42:784–791

Brown R, Colter N, Corsellis JAN et al (1986) Postmortem evidence of structural brain changes in schizophrenia. Differences in brain weight, temporal horn area, and parahippocampal gyrus compared with affective disorder. Arch Gen Psychiatry 43:36–42

Bruton CJ, Crow TJ, Frith CD, Johnstone EC, Owens DG, Roberts GW (1990) Schizophrenia and the brain: a prospective clinico-neuropathological study. Psychol Med 20:285–304

Buss DM, 50 coauthors (1990) International preferences in selecting mates: a study of 37 cultures. J Cross-Cultural Psychol 21:5–47

Crow TJ, Ball J, Bloom SR et al (1989a) Schizophrenia as an anomaly of development of cerebral asymmetry. A postmortem study and a proposal concerning the genetic basis of the disease. Arch Gen Psychiatry 46:1145–1150

Crow TJ, Colter N, Frith CD, Johnstone EC, Owens DGC (1989b) Developmental arrest of cerebral asymmetries in early onset schizophrenia. Psychiatry Res 29: 247–253

Crow TJ (1993) Sexual selection, Machiavellian intelligence and the origins of psychosis. Lancet 342:594–598

Daniel EG, Goldberg TE, Gibbons RD, Weinberger DR (1991) Lack of a bimodal distribution of ventricular size in schizophrenia: a Gaussian mixture analysis of 1056 cases and controls. Biol Psychiatry 30:887–903

Darwin C (1871) The Descent of man, and selection in relation to sex (facsimile of original published by J. Murray in 1981, London). Princeton University Press, Princeton

Dunlap CB (1924) Dementia praecox: some preliminary observations on brains from carefully selected cases and a consideration of certain sources of error. Am J Psychiatry 80:403–421

Eaton WW, Mortenson PB, Herrman H et al (1992) Long-term course of hospitalisation for schizophrenia. I. Risk for hospitalisation. Schizophr Bull 18:217–228

Falkai P, Bogerts B, Greve B et al (1992) Loss of Sylvian fissure asymmetry in schizophrenia. A quantitative post-mortem study. Schizophr Res 7:23–32

Harvey I, Ron MA, Du Boulay S, Wicks D, Lewis SW, Murray RM (1993) Reduction of cortical volume in schizophrenia on magnetic resonance imaging. Psychol Med 23:591–604

Jablensky A, Sartorius N, Ernberg G et al (1992) Schizophrenia: manifestations, incidence and course in different cultures. A World Health Organization ten country study. Psychol Med [Suppl] 20:1–97

Pakkenberg B (1987) Postmortem study of chronic schizophrenic brains. Br J Psychiatry 151:744–752

Schneider K (1974) Primare und sekundäre Symptome bei der Schizophrenie (Fortschritte der Neurologie und Psychiatrie 1957; 25:487–490). Reprinted in Themes and Variations in European Psychiatry, edited by SR Hirsch, M Shepherd and translated by H Marshall, Bristol: J. Wright. pp 40–44

Zipursky RB, Lim KO, Sullivan EV, Brown BW, Pfefferbaum A (1992) Widespread cerebral grey matter volume deficits in schizophrenia. Arch Gen Psychiatry 49: 195–205

Subject Index

Springer-Verlag and the Environment

We at Springer-Verlag firmly believe that an international science publisher has a special obligation to the environment, and our corporate policies consistently reflect this conviction.

We also expect our business partners – paper mills, printers, packaging manufacturers, etc. – to commit themselves to using environmentally friendly materials and production processes.

The paper in this book is made from low- or no-chlorine pulp and is acid free, in conformance with international standards for paper permanency.

MIX
Papier aus verantwortungsvollen Quellen
Paper from responsible sources
FSC® C105338

If you have any concerns about our products,
you can contact us on
ProductSafety@springernature.com

In case Publisher is established outside the EU,
the EU authorized representative is:
Springer Nature Customer Service Center GmbH
Europaplatz 3, 69115 Heidelberg, Germany

Printed by Libri Plureos GmbH
in Hamburg, Germany